Resilience in the English Small-Scale Fishery

AF621044

Rebecca Korda • Tim Gray • Selina M. Stead

Resilience in the English Small-Scale Fishery

Small Fry but Big Issue

Springer

Rebecca Korda
Newcastle University
Brighton, UK

Tim Gray
Newcastle University
Northumberland, UK

Selina M. Stead
University of Stirling
Stirling, UK

ISBN 978-3-030-54244-3 ISBN 978-3-030-54245-0 (eBook)
https://doi.org/10.1007/978-3-030-54245-0

© The Editor(s) (if applicable) and The Author(s), under exclusive license to Springer Nature Switzerland AG 2021
This work is subject to copyright. All rights are reserved by the Publisher, whether the whole or part of the material is concerned, specifically the rights of translation, reprinting, reuse of illustrations, recitation, broadcasting, reproduction on microfilms or in any other physical way, and transmission or information storage and retrieval, electronic adaptation, computer software, or by similar or dissimilar methodology now known or hereafter developed.
The use of general descriptive names, registered names, trademarks, service marks, etc. in this publication does not imply, even in the absence of a specific statement, that such names are exempt from the relevant protective laws and regulations and therefore free for general use.
The publisher, the authors, and the editors are safe to assume that the advice and information in this book are believed to be true and accurate at the date of publication. Neither the publisher nor the authors or the editors give a warranty, expressed or implied, with respect to the material contained herein or for any errors or omissions that may have been made. The publisher remains neutral with regard to jurisdictional claims in published maps and institutional affiliations.

This Springer imprint is published by the registered company Springer Nature Switzerland AG
The registered company address is: Gewerbestrasse 11, 6330 Cham, Switzerland

This book is dedicated to Becca's partner, Iain, and their son, Toby, as well as to the inshore fishers living and working around the English coast. You are all inspirational.

Acknowledgements

This book originated from a PhD thesis submitted by Rebecca Korda to Newcastle University in 2019, supervised by Professors Tim Gray and Selina Stead. We are grateful to the university for providing the facilities which made the PhD doctoral research possible and to the examiners (Professors Svein Jentoft and Jeremy Phillipson) whose feedback helped shape this work. Becca thanks her employer, Natural England, for granting several periods of leave to undertake fieldwork and write up her thesis though Natural England is in no way responsible for any of the views expressed in this book. We would like to wholeheartedly thank all of the respondents, for their graciousness in giving their time and entrusting us with their experiences. Thank you for being such kind and generous hosts. Becca extends thanks to B. Pool who accompanied her on some of her travels and to K. Cook, L. Paver and J. Selvage for putting her up during her tour of the country. Thanks also go out to J. Percy, C. Williams, B. Pool, H. Burrell and G. Ellis for their invaluable advice and feedback.

Finally, Becca acknowledges the enormous support she received from her family and friends. She thanks her son Toby, who joined her halfway through her studies, for bringing such joy and inspiration and sustaining her with mud pics. She thanks her partner, Iain, for supporting and encouraging her during the last 6 years, as well as picking up the lion's share of parenting duties to provide her with the space and time to focus on this work. She thanks her parents for their enthusiasm, sureness, endless cups of tea and for cheerily helping out with child care. She also wishes to thank her brother Ben for being an eternal cheerleader. Finally, special thanks go to her friends, T. Rich, S. Colover, L. Broom, O. Stephens-Denyer, K. Spears, K. Franke, R. Mayers, B. Hicks, J. Khalfan, R. Armstrong, L. Torres, F. Wood, P. Shackley, H. Burrell, S. Bostock, T. Bartlett, J. Daniels, C. Williams and D. Kirk-Adams, with more gratitude and affection than can well be put down here.

Abbreviations

CBFM	Community-based fisheries management
Cefas	Centre for Environment, Fisheries and Aquaculture Science
CFO	Chief Fisheries Officer [full title Chief Inshore Fisheries and Conservation Officer]
CFP	Common Fisheries Policy of the European Union
Defra	Department for Environment, Food and Rural Affairs
EA	Environment Agency
EEZ	Exclusive economic zone
EFZ	Exclusive fishing zone
EU	European Union
FA	Fishermen's Association
FAO	Food and Agriculture Organization of the United Nations
FGD	Focus group discussion
FLAG	Fisheries Local Action Group
FLC	Fish Locally Collaborative
FQA	Fixed quota allocation
ICES	International Council for the Exploration of the Sea
ICSF	International Collective in Support of Fishworkers
IFCA	Inshore Fisheries and Conservation Authority
ITQs	Individual transferable quotas
IUU	Illegal, unreported and unregulated fishing
KI	Key informant
LIFE	Low Impact Fishers of Europe
MAFF	Ministry of Agriculture, Fisheries and Food
MARE	Centre for Maritime Research
MCCA	Marine and Coastal Access Act 2009
MCZ	Marine Conservation Zone
MPA	Marine protected area
MSC	Marine Stewardship Council
MSY	Maximum sustainable yield
NEF	New Economics Foundation

NFFO	National Federation of Fishermen's Organisations
NGO	Non-governmental organisation
NUTFA	New Under Ten Fishermens Association
PO	Producer Organisation
RBS	Registration of Buyers and Sellers
SFC	Sea Fisheries Committee
SFF	Scottish Fishermen's Federation
SLA	Sustainable livelihoods approach
SSF	Small-scale fisheries/fishers
TAC	Total allowable catch
TBTI	Too Big to Ignore
UKIP	United Kingdom Independence Party
UNDP	United Nations Development Programme
VMS	Vessel monitoring system
WFFP	World Forum of Fisher People
WWII	World War II

Contents

Chapter 1
Introduction

> *'Small-scale fisheries are too big to ignore and too important to fail' (Ratana Chuenpagdee)*
> *'They see us as such small fry that we can just be brushed aside' (KI-59)*

1.1 Marginalisation of Small-Scale Fisheries (SSF)

This book is intended as a contribution to our understanding of the uncertain situation of small-scale fisheries (SSF) which face marginalisation in most coastal countries, and possible extinction in some. This situation is paradoxical in that SSF constitute by far the largest proportion of commercial marine fishers in the world yet possess the least power by which to ensure the continuation of sustainable livelihoods. In this introductory chapter, we explain why SSF are so marginalised; how there has been a powerful backlash against this marginalisation during the last 30 years; what are the main ideational currents supporting this backlash; and what is the enduring value of SSF that justifies that support. But first, we must deal with the vexed question of what defines small-scale fisheries.

There is considerable controversy over what constitutes a small-scale fishery (SSF). For one thing, many different names are given to SSF, including artisanal, subsistence, inshore, coastal, traditional, and low-tech. For another thing, there is no single internationally-accepted definition for SSF (Jentoft and Eide 2011; García-Flórez et al. 2014; Jacinto and Pomeroy 2011; Basurto et al. 2017; Davies et al. 2018). Smith and Basurto (2019) point out that definitions of SSF are sometimes chosen to serve management purposes. It may be that the best way to explain the meaning of SSF is to contrast the above names with their antonyms: small-scale versus large-scale; artisanal versus industrial; subsistence versus commercial; inshore versus offshore; coastal versus distant; traditional versus modern; and low-tech versus high-tech (McConney and Charles 2012). However, none of these characterisations definitively describes SSFs, because each SSF is unique (Johnson and

© The Editor(s) (if applicable) and The Author(s), under exclusive license to Springer Nature Switzerland AG 2021

R. Korda et al., *Resilience in the English Small-Scale Fishery*,
https://doi.org/10.1007/978-3-030-54245-0_1

Pálsson 2015), and can only be fully understood in the specific context of its own particular situations. As McConney and Charles (2012) point out, what constitutes small-scale in one situation may constitute large-scale in another situation: for example, in the English fleet, small-scale signifies vessels measuring 10 m or under, whereas in North America, most small-scale vessels are 15–20 m in length. We will return to this issue of the definition of SSF in Chap. 3: in the meantime, we will interpret SSF as meaning inshore, coastal fishing.

Small-scale fisheries (however defined) have not always been marginalised. Indeed, the very origin of fishing was on a small-scale. It is only with the advent of industrialisation in the nineteenth century and globalisation in the twentieth century that SSF came under the threat of displacement by larger and more efficient vessels. Kolding et al. (2014, pp. 1, 2) point out that SSF have a much longer history than industrial fisheries, yet have now been side-lined by them:

> Small-scale or artisanal fisheries date back to the dawn of human history, and still constitute a widespread activity and occupation. Large-scale or industrial fisheries on the other hand have evolved during the past century and have now come to dominate the realms of policy and science…Modern industrial fisheries have captured the limelight and shaped most fisheries paradigms, relegating the traditional, 'primitive' and 'inefficient' activities that habitually characterize SSFs to the margins.

One illustration of SSF's marginal status is the fact that that although 50 million of the global total of 51 million fishers are small-scale (Salmi 2015), SSF have until comparatively recently not been studied nearly as much as large-scale fisheries (Jentoft et al. 2017a; Lloret et al. 2018; Salas et al. 2007, 2019; Chuenpagdee 2011b; Finkbeiner et al. 2015; Kolding and van Zwieten 2011; Islam and Berkes 2016). For example, Percy said that of the 3924 scientific papers that have been published on discards, 3760 were about large-scale fisheries and only 164 were about SSF (Oliver 2019g). As Kolding et al. (2014) note, one reason for this historical lack of research on SSF is that they are perceived to be inefficient and uneconomic (see also Pinkerton and Davis 2015; Ifremer 2007; Krogseng 2016). The assumption is that by replacing SSF with larger industrial vessels, productivity would be improved and economic returns would rise (Cohen et al. 2019). Commentators report that for some developmental economists, this is the price that must be paid for progress and modernisation: it is an inevitability that SSF will eventually be superseded (Jentoft and Chuenpagdee 2018; Pauly 2011; Jentoft et al. 2017b).

> The popular paradigm during the development decades (1950–1970s) was that the natural progression of the world's fishing was necessarily towards the industrial mode. Nations worldwide promoted this mode of fisheries development with strategies focusing almost exclusively on large-scale fisheries and the need to increase fishing effort and capacity…it was assumed that the [SSF] subsector would either expand its scale of production by adopting large-scale fishing techniques or else provide labour to industrial operations and gradually disappear (Carvalho et al. 2011, p. 360).

Haakonsen (1992, p. 33) traces this doom-laden scenario to the late nineteenth century:

> The fisheries sector['s]…industrialisation in Europe and North America can be said to have started in the second half of the nineteenth century with the introduction of steam engines on

> fishing vessels which in turn allowed for the installation of winches, refrigeration systems, etc., culminating in some of today's giant factory ships with all the latest navigation, fish detection, processing and preservation technologies. It is not surprising then that the industrialisation-equals-progress belief was readily accepted as a guide for developing the often marginally exploited fisheries sector in the new emerging nations in Africa in the 1950s and 1960s. Whatever fisheries existed they were, in their very basic artisanal form, seen as backward and inefficient and bound to disappear over time once the industrial part of the sector 'took off', to use a prevailing development terminology from the 1960s.

Béné et al. (2015a, p. 19) point out that some commentators portrayed SSF as "backward, informal and marginal economic actors that were doomed to disappear with economic development and modernization".

This narrative of inevitable decline of SSF has been underpinned by two powerful interconnected economic concepts: neo-liberalism and globalisation. Neo-liberalism, the ideology of capitalism which advocates deregulation, lower government spending, and reduced taxes, prioritises the private sector, promotes the free market, and endorses private property rights (Pinkerton 2017), enjoyed widespread support after the collapse of the Soviet bloc's command and control regimes in the 1990s. Knutson (2017) names this neo-liberal turn 'corporate domination'; Knott and Neiss (2017) describe its processes of 'privatization', 'marketization' and 'financialization'; Longo and Clark (2012, p. 204) refer to its "high-tech, capital-intensive methods"; and Lalancette (2017, p. 47) mentions "neoliberal principles of market governance, commodification of natural resources, profit-maximization, individualization and property rights".

Applying neo-liberalism to the fishing industry, fisheries economists from the neo-liberal perspective argue that large-scale fisheries are far more efficient economically than are SSF because of economies of scale, and they drive down the price of fish to consumers. Macfadyen et al. (2011, p. 78) comment on EU (European Union) fisheries that "the economic data presented here show that...the economy is not well served by SSF...Average earnings per fisherman employed in SSF are significantly lower than in large-scale fisheries...the value of landings from large-scale fisheries is almost three times higher than from SSF" (see also Ifremer 2007). Hardin's famous notion of the 'tragedy of the commons' (Hardin 1968, p. 1244) adds a Malthusian strand to the neo-liberal critique of SSF, arguing that free access to coastal marine resources is a recipe for disaster as the number of artisanal fishers increases to the point at which fish stocks collapse because of overfishing: "Freedom in a commons brings ruin to all". Like Hardin, Gordon (1954) asserts that property rights of some kind are necessary to prevent excessive exploitation of fish stocks. One important manifestation of this neo-liberal turn is the adoption of 'catch shares' systems, particularly in the form of individual transferable quotas (ITQs). Bodwitch (2017, p. 89) reports that "Starting in the late 1970s, fishery economists argued that privatization of fishing rights, in the form of individual transferable quota, could stop overfishing" (see also Bennett 2017).

Closely linked to neo-liberalism, globalisation is the second economic concept that seemed to seal the fate of SSF. Defined as the limitless expansion of movement across national boundaries of goods, services, labour, capital, technology, and data, globalisation has led to increasing integration of regional, national and local economies throughout the world, offering vast opportunities for economic development. Meyer (2017, p. 80) says "globalization has been a major source of economic growth and prosperity", while Warner (2005) claims globalisation has created millions of new jobs, and Bhagwati (2004) asserts that economic globalisation is the only way to combat global poverty. Daboub and Calton (2002, p. 1) describe globalisation as "the most important development of our time", while Warner (2005, p. 238) claims that in the mid-1990s, "economists and politicians everywhere were proclaiming the dawn of a new age for humanity…'globalisation' was widely accepted as the new world order". Hines (2000, p. vii) quotes world leaders who proclaim that globalisation is a fact of life: "Globalisation is not a policy choice, it is a fact" (Bill Clinton); "Globalisation is 'irreversible and irresistible'" (Tony Blair). Many writers argue that the march of globalisation is relentless, and that governments who try to resist it, will, like Canute, fail badly, and in doing so, undermine the sustainability of their entire economies. On this view, economic development or 'progress' lies inexorably in the adage 'large is beautiful'.

Some commentators have added an evolutionary interpretation to the march of globalisation. For example, Jentoft (2019, pp. 313–314) characterises globalisation in Herbert Spencer's evolutionary terms of the 'survival of the fittest':

> Spencer is famous for the term 'survival of the fittest', where, like in nature, social evolution rids us of things that are not well adapted. Policy-makers and governors may be tempted to look at small-scale fisheries development in a similar way. As a natural process, the idea would be that the industrial, large-scale fisheries as a more efficient mode, would unavoidably supplant small-scale fisheries. Any effort to save small-scale fisheries from becoming extinct is, at best, delaying their demise because it would be against the 'the law of nature'. Small-scale fisheries are bound to perish, as an adaptive process tantamount with evolution.

So it has been argued that SSF are the past, and industrial fisheries are the future: the economies of scale cannot and must not be ignored. Advocates of large-scale fisheries point out that industrial fishing has contributed hugely to the enormous expansion of global production of caught fish—from 18 million tonnes to 90 million tonnes during the last 50 years (Eide et al. 2011). Song et al. (2018) report that 4% of fishers produce 76% of the global catch, and during 2007–2012, the 16 largest international fishing corporations nearly doubled their revenues. There are many examples around the world of this shift from small to large fisheries. For instance, Bavinck (2011) describes the occurrence of industrial fishing in post-independence India as the 'Blue Revolution' in parallel to the 'Green Revolution' of the industrialisation of agriculture. Monnereau and McConney (2015, p. 224) refer to "the industrialization of the world's oceans" (see also Smith 2000). Longo and Clark (2012) report how industrialised fishing for blue fin tuna in the Mediterranean Sea threatens to wipe out SSF tuna trap fishing. Referring to Greenland's coastal halibut fishery, Jacobsen (2013, p. 16) says "the dominant development discourse in Greenlandic fisheries governance…asserts that big is simply better". Jacobsen (2013) says another reason why fisheries managers prefer large-scale vessels to small-scale vessels is because SSF are regarded as much harder to control: indeed, there is a question over whether SSF is ungovernable.

Given these understandings, it is hardly surprising that in the name of 'development' and 'modernisation' many governments have been encouraging artisanal fishers to move away from SSF to large–scale industrial fisheries (Jacobsen 2013; Kraan 2011). Brattland (2014, p. 3) says in Norway, this has been dubbed a process of 'cyborgization', whereby the close 'organic' relationship between SSF and their vessels is replaced by a detached 'mechanical' and electronic relationship: "small-scale coastal fishing vessels are transformed into technologically sophisticated killing-fish machines or 'fishing cyborgs'". Lalancette (2017, p. 52) describes it as "professionalization…Fishers are being redefined as 'business managers' with efficiency and profitability as the primary goals". Campling et al. (2012, p. 182) refer to it as a process of commodification of fishing: "fish have changed from being produced as food for producing communities to being produced as commodities for sale". Bailey (2018) describes how across the world, large-scale fishers have long been trying to force SSF out of business. De Schutter (2012) claims that "the encroachment of industrial fleets…poses a major existential threat to these traditional fishing communities".

Several writers claim that SSFs are already insignificant. For example, de Vos and Kraan (2015, p. 629) say "many small-scale métiers…are out of sight, as they are outside the bounds of data collected", and refer to the "relative invisibility of the small-scale fisheries in the Netherlands" (see also Acott et al. 2018). Arias-Schreiber et al. (2019) claim that SSF in the Baltic Sea is rapidly disappearing. Seixas et al. (2019) report that in Brazil, during 1962–1989, SSF landings as a proportion of total landings shrunk from 80% to 20% as the government strove to develop the industrial fleet (see also De Mattos and Wojciechowski 2019). Pauly (2006, p. 16) warns that SSF is in danger of extinction: "In the long term (two to three decades?), fisheries and fishing-based cultures will not survive if we do not manage to put small-scale fisheries and resources first".

1.2 Backlash Against the Marginalisation of SSF

However, during the last 30 years there has been a concerted and forceful backlash against this marginalisation of SSF. One element of this backlash is the claim that in some countries, especially developing countries, far from declining, SSFs have been growing. For example, Haakonsen (1992, pp. 33, 47–48) argues that in some West African countries, SSF have adopted simple technological improvements to flourish where industrial fisheries have failed:

> In West Africa, only in a couple of countries did industrial fisheries 'take off', and then only to a moderate extent and with a limited degree of success. The 'backward' artisanal fisheries, on the other hand, has prevailed, expanded and even prospered by comparison, adopting simple, but efficient technological innovations…the reason for the progress of much of the artisanal fisheries in countries like Senegal and Ghana has been the fishermen's adaptability

> and readiness to incorporate new technologies, thus proving themselves quite different from the image of the backward, narrow-minded and ultratraditional 'peasant-type. First and foremost has been the rapid acceptance of the outboard engine on traditional canoes, which took place in the late 1950s in both countries.

Likewise, Jul-Larsen (1992), argues that SSF in West African countries can improve economically from within their own traditional systems by adopting some technological innovations without embracing the capitalist system of industrial fisheries. As we shall see, this argument chimes with Schumacher's advocacy of alternative or intermediate technology. Kolding et al. (2014) point out that far from dwindling, SSF is growing (at least in developing countries) with more people joining it and introducing improvements in its fishing technology.

Another element of the backlash is the recent huge upsurge of interest in SSF across the world because of a claim that they are much more sustainable than is industrial fishing. As Carvalho et al. (2011, p. 360) noted, a global crisis of overfishing has begun to undermine much support for industrial methods:

> With the crises in world fisheries, the industrial model of development has increasingly been put under scrutiny. After more than half a century of a strong modernisation imperative that put economic efficiency high on the policy agenda for fisheries, the policy arena is finally becoming more conducive to sustaining small-scale fisheries. The notion that small-scale fisheries are probably our best option for a sustainable use of fisheries resources, assembling most of the criteria required for an enlightened fisheries policy in terms of employment, income distribution, energy consumption, and product quality, has gained significance…with many studies emphasising the social significance, cultural diversity and economic importance of sustaining this subsector.

Armitage et al. (2017a) claim that the need to sustain local coastal fishing communities pursuing fishing on a small scale basis, is now being acknowledged as a global priority, and Allison and Ellis (2001) state that the importance of sustaining SSF is becoming increasingly recognized by development organisations and fisheries managers. Béné et al. (2015a, p. 14) state that "At the international level, substantial progress has been made in recent decades to raise the profile of small-scale fisheries". For example, in 2014 the Food and Agriculture Organization (FAO) of the United Nations issued the FAO Voluntary Guidelines for Securing Small-Scale Fisheries in the Context of Food Security and Poverty Alleviation, which was the first internationally agreed instrument dedicated exclusively to the SSF, asserted that small-scale fisheries must have secure tenure rights to the fishery resources that sustain their livelihoods, their cultural well-being, and their sustainable development:

> States should, where appropriate, grant preferential access of small-scale fisheries to fish in waters under national jurisdiction, with a view to achieving equitable outcomes for different groups of people, in particular vulnerable groups. Where appropriate, specific measures, inter alia, the creation and enforcement of exclusive zones for small-scale fisheries, should be considered. Small-scale fisheries should be given due consideration before agreements on resource access are entered into with third countries and third parties" (FAO 2015, p. 6, para 5.7).

This was seen as a "historic moment" and "potential turning point" (Jentoft et al. 2017b, p. 3) and "a milestone event for small-scale fisheries worldwide" (Franz and

Barragán-Paladines 2017, p. 36. There are many examples across the world of projects pushing back against industrial fisheries in recent years. Ratner and Allison (2012, p. 372) state that "The critique of modernisation and structural adjustment policies, and recognition of their limitations and undesirable effects on the poorest and most vulnerable, gave rise to a series of projects in the 1990s and 2000s that focused on strengthening fisherfolk's livelihoods". For example, in Brazil, Gasalla (2011, p. 189) describes a process of "re-artisanalization" of fishing activities in coastal areas following a decline and collapse of industrial fisheries. Aryeetey (2002, p. 336) says "failure of the newly established industrial fleets to deliver the expected outputs in many African countries has made governments turn attention, once again, to the activities of the artisanal sector". In South Africa, Sowman (2006, p. 60) reports that the Marine Living Resources Act in 1998 "legally recognised subsistence fishers and made provision for the declaration of coastal areas for their exclusive use. In 2001, a limited [SSF] commercial fisheries sector was created. These changes indicated government's commitment to addressing the historical marginalisation of small-scale fishers". This policy marked "a paradigm shift" for the rehabilitation of the traditional black SSF sector in the country which had been systematically discriminated against during the apartheid era in favour of the white industrial sector: "This policy aims to provide redress and recognition to the rights of small-scale fisher communities in South Africa previously marginalised and discriminated against in terms of racially exclusionary laws and policies, individualised permit-based systems of resource allocation and insensitive impositions of conservation-driven regulation" (DAFF 2012, p. 17, 1). Islam and Berkes (2016, p. 2) report that "Canadian courts have established that...subsistence fisheries of indigenous people have priority over all other uses of the resource".

Moreover, several associations have been established to represent SSFs at national and international levels and monitor their treatment by governments and intergovernmental authorities such as the EU. For example, in the Mediterranean and Black Sea region, a digital platform has been set up called the Friends of Small-Scale Fishermen, partly funded by the EU, to provide a mapping tool to monitor projects in SSF and improve cooperation in such projects. Another association is the World Forum of Fish Harvesters and Fish Workers, which is a civil society organisation based in Uganda established to bring together SSFs from across the world to discuss key issues facing them. It was formed in the late 1990s as a response to SSFs' exclusion from international bodies such as the FAO, and has held five general assembly meetings since 1997 (Basurto et al. 2017). A similar association is the World Forum of Fisher Peoples which focuses especially on providing capacity building for SSF organisations.

Furthermore, many international bodies have endorsed the case for supporting SSF. Basurto et al. (2017) reports that funding for projects relevant to SSF across the globe during 2000-2016 was a total of $1.8 billion, most of which came from multilateral aid agencies, especially the World Bank. Such initiatives are characterised by Penca (2019) as 'transnational localism'—i.e. the use of international mechanisms to champion local SSF. Penca points to three major steps in this direction: the EU's reformed Common Fisheries Policy (CFP) adopted in 2013 included a commitment to support SSF; the FAO's Voluntary Guidelines for Securing Sustainable Small-Scale Fisheries, published in 2014; and the adoption of the Sustainable Development Goals in 2015, especially SDG 14 which urges governments to provide access to marine resources and markets to SSF. In 2018, the Ministerial Council of the Caribbean Regional Fisheries Mechanism endorsed a protocol on Sustaining Sustainable Small-Scale Fisheries in the region, where SSF accounts for 95% of fisheries (CRFM 2018). In Malta in 2018, a Ministerial Declaration on a Regional Plan of Action for Small-Scale Fisheries in the Mediterranean and the Black Sea was signed by Karmenu Vella (then EU Commissioner for the Environment, Fisheries and Maritime Affairs) at a High Level Conference which "reinforces opportunities for small-scale fishermen in the Mediterranean and the Black Sea regions by giving them a voice in the decisions that affect their livelihoods. It aims at enhancing their capacities in contributing to food security and in achieving economic, social and employment benefits while safeguarding environmentally sustainable fishing practices" (EU Commission 2018). A digital platform 'Friends of Small-scale fishermen' was also launched, funded in part by the EU, to record projects and investment in SSF in the region, and encourage cooperation between them. In 2016, the Commission for Natural Resources of the European Parliament's powerful Committee of the Regions urged the EU to take measures to prevent the further decline of SSF, declaring SSF "a small but indispensable part of the local economy" (EU Parliament 2016). Despite much criticism of the EU for subsidising the expansion of the industrial sector, it has been pointed out that

> Under the Common Fisheries Policy, the fleet has preferential access in the 12-nautical-miles coastal band of the EU and is exempted from a number of obligations that apply to larger vessels, such as…fishing authorisations, landing declarations, sales notes, separate storage…the exemption from reporting catches under 50kg...is tantamount to a general exemption from logbook reporting. More 'affirmative action' comes from the European Maritime and Fisheries Fund (MAF 2016).

In 2014, the reformed CFP referred to the need for a fair standard of living for the fisheries sector (including SSF) in EU waters, and for preferential access to fishing opportunities for SSF, including incentives to use fishing methods with low impacts on fisheries resources and marine ecosystems, while the European Maritime and Fisheries Fund required Member States which had more than 1000 small-scale fishing vessels to produce action plans for their development, competitiveness and sustainability (Percy 2016, 2015). In a 2011 report commissioned by the European Parliament, it is stated "The maintenance of small-scale fleets is a widespread policy objective in many EU Member States…Small-scale fishers can be considered as the '*guardians of the coastal zone*', similar to the role of some farmers in the rural areas" (Macfadyen et al. 2011, p. 13, 81) [italics in original]. In 2012, an international group of SSF founded an advocacy group of local SSF organisations called Low Impact Fishers of Europe (LIFE) to give a voice to the thousands of small boat fishers across Europe (Penca 2019). Jerry Percy, then executive director of LIFE, explained that "LIFE is here to give them a clear and coherent voice at the political heart of Europe…there is a growing recognition that it is now vital to have a dedicated voice to champion their cause" (Oliver 2014). By 2016, LIFE had 7000 active inshore fisher members (Percy 2016). Percy (2015) also drew attention to Defra (Department for Environment, Food and Rural Affairs)'s 2027 Vision Document in which it states that

> Access to fisheries continues to be available to small-scale fisheries vessels, even if in some cases that is not the most economically efficient way of harvesting the resource. This is because the wider economic, social and environmental benefits of small-scale fishing can outweigh the comparative inefficiency in harvesting the resource and make a significant economic and social contribution to the lives of individuals and coastal communities.

1.3 Ideational Currents Supporting the Backlash Against Marginalisation of SSF

This backlash in favour of a return to small-scale fisheries is in part inspired by Schumacher's (1973) book entitled *Small is beautiful: A study of economics as if people mattered*, which has united several strands of activism against western neo-liberalism. Schumacher (1973, section 1) declares that "the whole world is now in a process of westernisation" but argues that western capitalism is unsustainable, over-exploiting and polluting non-renewable natural resources, and that exporting western technology to the developing world (technology transfer) would merely replicate that unsustainability: a case of "the bland leading the blind". His philosophy of sufficiency urges communities to adopt appropriate or intermediate technologies and seek to maximise well-being, not commodity production—i.e. to pursue human worth and dignity, not soul-destroying material affluence. On smallness, Schumacher (1973, sections 18, 2, 3, and 5) says:

> In general, small enterprises are to be preferred to large ones…Small-scale operations, no matter how numerous, are always less likely to be harmful to the natural environment than large-scale ones…There is wisdom in smallness if only on account of the smallness and patchiness of human knowledge, which relies on experiment far more than on understanding…Today, we suffer from an almost universal idolatry of gigantism. It is therefore necessary to insist on the virtues of smallness.

On technology, Schumacher (1973, sections 10 and 12) says we need

> a different kind of technology, a technology with a human face, which instead of making human hands and brains redundant, helps them to become far more productive than they have ever been before. As Gandhi said, the poor of the world cannot be helped by mass production, only by production by the masses…I have named it intermediate technology to signify that it is vastly superior to the primitive technology of bygone ages but at the same time much simpler, cheaper, and freer than the super-technology of the rich. One can also call it self-help technology, or democratic or people's technology - a technology to which everybody can gain admittance and which is not reserved to those already rich and powerful…intermediate technology will be 'labour-intensive' and will lend itself to use in small-scale establishments.

For the neo-liberal economist, "growth of GNP [gross national product] must be a good thing, irrespective of what has grown and who, if anyone, has benefited. The idea that there could be pathological growth, unhealthy growth. disruptive or destructive growth, is to him a perverse idea" (Schumacher 1973, section 3). For the neo-liberal economist, economic growth depends on increased output (what the fisheries economist would characterise as increased catch per unit of effort (CPUE)). But for Schumacher (1973, section 12), jobs are more important than output: "It is more important that everybody should produce something than that a few people should each produce a great deal…Within manufacturing, there should be imaginative exploration of small-scale, more decentralised, more labour-using forms of organisation".

Schumacher's philosophy feeds off and reinforces the anti-economic growth movement that began in 1972 with the publication of the report entitled *The Limits to Growth* by Donella Meadows, Dennis Meadows, Jørgen Randers, and William Behrens which was based on a computer simulation of unsustainable endless economic and population growth with a finite supply of resources. The Brundtland Report, *Our Common Future,* published in 1987, derived the concept of 'sustainable development' to restrain economic growth to the extent that it left sufficient natural resources for succeeding generations, and this was the foundational principle for the Rio Earth Summit on Sustainable Development in 1992. More recently, the term 'unjust uneconomic growth' has been used to characterise the damaging nature of capitalist excess (Sabau and van Zyll de Jong 2015).

Another ideational current supporting the backlash against the marginalisation of SSF is the anti-globalisation movement. During the twenty-first century, there has been growing disillusion with globalisation, which has been blamed for the Asian financial crisis in 1997 and the western financial crisis in 2008, that ushered in an age of austerity for many developing and developed nations. Moreover, as Daboub and Calton (2002, p. 1) explain, globalisation has been "blamed for increasing the gap between rich and poor, accelerating the destruction of the environment, and threatening human rights". Held and McGrew (2003, pp. 29; 30) say critics argue that economic globalisation "is directly responsible for widening disparities in life chances across the globe—a deepening polarisation of income and wealth…the segmentation of the global workforce into those who gain and those who lose…the growing marginalisation of the losers from the global economy…and the erosion of social solidarity…Unless economic globalisation is tamed, so the argument goes, a new barbarism will prevail as poverty, social exclusion and social conflict envelop the world". Hoffmann (2003, p. 108) says "Economic globalisation has…become a formidable cause of inequality among and within states". Woods (2003, p. 465) writes "Globalisation is cementing old inequalities between 'haves' and 'have-nots'". Even advocates of globalisation such as Meyer (2017, p. 80) admit its inegalitarian outcomes: "international trade has made people around the world better off *on average*. But that average hides the unequal nature of the gains from globalisation" [italics in original]. Warner (2005, p. 238) says "the accelerated pace of globalisation has…proven to be a destructive force to millions, perhaps billions, of people the world over. Some staggering statistics are illustrative: two billion people live on less than $2 a day, while three billion people live on less than $3 a day…In sub-Saharan Africa, per capita growth was 36 percent between 1960 and 1980, and then it collapsed completely under globalisation, actually falling 15% between 1980 and 1998. The gap between rich and poor continues to grow, as '458 billionaires possess more wealth than do half of humanity'".

Bleiker (2008, p. 5) says anti-globalisers reject the belief in the "inexorable irreversibility of free-market globalisation", and argue that "globalisation is not an inevitable phenomenon, but a constructed narrative, a 'political project which can be responded to politically'". Norberg-Hodge (2001, p. 179) argues there is nothing natural or inevitable about globalisation: "it is occurring because governments are actively promoting it and subsidising the framework necessary to support it". Hoffmann (2003, p. 108) says "globalisation is neither inevitable or irresistible. Rather, it is largely an American creation, rooted in the period after World War II and based on U.S. economic might". The UNDP [United Nations Development Programme] (2003, p. 429) claims "None of these pernicious trends—growing marginalisation, growing human insecurity, growing inequality—is inevitable. With political will and commitment in the global community, they can all be reversed". Anti-globalisers fall into two categories: moderates and radicals. As we shall see, the moderates enunciate a strategy of adaptive resilience, whereas the radicals enunciate a strategy of transformational resilience. The moderates seek not to reverse globalisation, but to control it in order to restrict its potential to increase inequality. As Stiglitz (2003, p. 481) put it, "Globalisation can be reshaped, and when it is, when it is properly, fairly run, with all countries having a voice in policies affecting them, there is a possibility that it will help create a new global economy in which growth is not only more sustainable and less volatile but the fruits of this growth are more equitable shared". UNDP (2003, p. 424, 429) argues that stronger governance at all levels is needed to tame globalisation: "The challenge of globalisation in the new century is not to stop the expansion of global markets. The challenge is to find the rules and institutions for stronger governance—local, national, regional and global—to preserve the advantages of global markets and competition, but also to provide enough space for human community and environmental resources to ensure that globalisation works for people—not just for profits".

The radical anti-globalisers reject the moderates' idea that globalisation can be tweaked or tamed as a largely futile effort "to lasso a tiger with cotton" (Hines 2000, p. ix). Instead, they argue that what is needed is to halt globalisation altogether and replace it with localism, closely related to the principles of decentralisation and subsidiarity. They want decision-making devolved to the lowest possible level, from central government to regional and local government who in turn devolve power to local communities which are free to conduct their own business in whatever way their residents choose, rather than have decisions made for them by centralised authorities who are ignorant of local needs (Jameson 2013). Hines (2000, p. 31) defines localisation as "something done by people, not something done to them", and he sees it (2000, p. 57) as a "commitment to a new paradigm in which communities can become agents rather than victims, with programmes that enable them to attack the structures of dependency and retake control of their destiny". Localists seek to minimize the damaging impact of globalisation by championing small-scale forms of economic and cultural activity—i.e. local self-sufficiency or autarky. Wills (2016) links localism to the communitarian tradition, freeing up and empowering local communities to act in their own interest. The Foundational Economy (a collective of European academic researchers who are challenging the orthodox top-down economic policy of maximising GDP), emphasises the importance of the bottom-up or local level as the foundation of a more meaningful existence: "the aim is citizen lives worth living…Start by finding what citizens and businesses want because only communities know what matters to them" (Foundational Economy 2018, p. 8, 11).

During the early 2000s, 90 organisations coalesced in the UK under the banner of 'Local Works' to promote localism (Boyle 2009). Localists claim that before the nineteenth century, the mode of production, distribution and exchange was primarily local, and it was only with the advent of the industrial revolution and imperialism that the local scale became disparaged. Advocates of localism sometimes characterise themselves as returning to that earlier stage of production—'relocalisation'. Localists point to the damage inflicted on the high street by large-scale retailers with their out-of-town sites and more recent internet shopping facilities sucking the heart out of local communities, and leaving 'dead town centres'. Boyle (2009, p. 15, 13) says that "centralism and giantism in business…[is] driving the 'ghost town' phenomenon, emptying communities of post offices, banks and pubs or transforming the places where we live into bland identikit centres where the economic and buying decisions are taken by distant purchasing executives, and where shopkeeper jobs have been replaced by low-paid check-out staff". He portrays this as an apocalyptic failure of our civic pride, and a surrender of our personal identity:

> It represents a slow emptying of our institutions, a destruction of our way of life as tangible as any terrorist attack, a rotting away of the nation of shopkeepers – of imagination, innovation and pride at local level – to a miserable slavish acceptance of whatever narrow aspects of life our centralised system choose to deliver. This is impoverishing our culture and the services and institutions we rely upon. It is also a far broader problem than the current debate about localism would have us believe. It represents a shift in our status as citizens, a diminution of our individual power, a reduction to the status of supplicants to giant and distant organisations…It is a debilitating sickness (Boyle 2009, p. 18).

Localists also draw on an environmental trope that local food production and consumption leave a lighter footprint on the earth, and avoid the polluting effects of industrial agriculture (McKibben 2010). Curtis (2003) calls this 'ecolocalism'. Another term is 'locavore', which refers to people who only eat food that is grown or produced within, typically, a 100-mile radius of their local community, compared to the average distance travelled by food products sold in supermarkets today of 1300–2000 mile. As locavores point out, at one time everyone was a locavore, before it became possible to transport food products long distances. Local produce is usually sold within 24 h of being harvested, and is claimed to taste better, be fresher, more nutritious, and healthier than supermarket food (which may be grown on factory farms and treated with pesticides and fertilisers), and gives support to farmers and small businesses in the community (Tep and Fauchald 2007).

In the UK, localism permeated the thinking of both left-wing and right-wing governments. On the left, Labour Party Prime Minister, Tony Blair (1997–2007) despite regarding globalisation as irresistible, associated his 'Third Way' initiative with 'new localism', attempting to devolve power to local authorities to regenerate local services while maintaining national standards of service provision (Coaffee 2005; Coaffee and Johnston 2005). On the right, Conservative Party Prime Minister David Cameron (2010–2016) under his initiative to shift from the 'Big State' to the 'Big Society' effectively created a Minister for Localism under the 2011 Localism Act, and conferred wide-ranging powers on local authorities (DCLG 2011). Eric Pickles, the Secretary of State for Communities and Local Government, said his priorities were localism, localism and localism. In 2014, two pressure groups, the National Coalition for Independent Action and Open Democracy launched a campaign in the UK called Localism Watch to support communities in their localism initiatives. More widely, writers such as Katz and Nowak (2018) say that a 'New Localism' is emerging in cities across the world including Preston, Barcelona, Valparaiso, Copenhagen, Ghent, Cincinnati, Dayton, Jackson, Los Angeles, Newark, Pittsburgh, San Antonio, Seattle, and St. Louis that are developing pragmatic, inventive and nimble post-capitalist solutions customized to their local problems. Wills (2016, p. 4) refers to the "localist turn".

However, localism has itself been criticised for the inequality it can foster. For example, Hinrichs (2003, p. 36, 37) claims that "local social interactions are not absent of intolerance and unequal power relations. Local communities and organizations may have chequered histories, replete with provincial bias and social exclusion…localisation becomes elitist and reactionary, appealing to narrow nativist sentiments". A related criticism is that it essentialises and romanticises the local (Mohan and Stokke 2000), turning a blind eye to its bigotry and narrow-mindedness. Wills (2016, p. 3) rehearses the charge that localism is "dangerously parochial, often associated with nostalgia and sentimentalism". In the UK, a Commission on the Future of Localism set up in 2017 with Lord Kerslake (a former head of the civil service and President of the Local Government Association) as chair, found in 2018 that the promise of localism had not materialised because of lack of support from the government: "the 'fundamental shift of power' promised by the Localism Act 2011 has not yet been achieved. To unlock the power of community, we need radical action that strengthens our local institutions; devolves tangible power, resources and control; ensures equality in community participation; and delivers the culture change required to enable local initiatives to thrive" (Locality 2018, p. 6; see also Stanton 2014). Both the advocacy of localism and the criticisms of it resonate with the situation of the English SSF, which exemplifies the values of community, but also reveals fissures and inequalities in some local areas.

1.4 Explosion of Research on SSF

In the academic study of fisheries, this backlash against industrialisation and globalisation has taken the form of an explosion of research on SSF. Smith and Basurto (2019, p. 1) refer to "the rising tide of interest buoying SSF", and state that "in recent decades, attention to SSF is on the rise, marked by a proliferation of scientific publications, the emergence of new global policy tools devoted to the small-scale sector, and concerted efforts to tally the size and impacts of the SSF on a global scale". Basurto et al. (2017) report that in 2016 there were 427 researchers on SSF from 62 countries Some of these researchers claim SSF are more resilient than their neo-liberal critics think. For example, Eide et al. (2011, p. 21) write: "Small-scale fisheries have defied many predictions just by staying around, and it is likely that they will be present for many years to come". Many of these studies have been linked to global conferences such as the World Small-Scale Fisheries Congress, which held its first meeting in 2010 in Thailand and led to the creation of the 'Too Big to Ignore' (TBTI) global network of researchers devoted to examining the challenges facing the SSF across the world (Chuenpagdee 2011b). From this network, hundreds of studies of SSF in scores of countries have been carried out and published in academic journals and edited books. In 2017, under the TBTI global partnership, a 858-page book was published, entitled *The small-scale fisheries guidelines: Global implementation,* which provided detailed advice for national and international policy-makers on steps to take to support the sustainable development of small-scale fisheries and their communities. Through this partnership, an SSF information system has been established with the intention of gathering data on SSFs around the world, which can then be shared with researchers and practitioners to aid understanding and help decision-makers to formulate better-informed policies (Basurto et al. 2017). In addition, the TBTI has set up a programme to provide training modules for professional development and life-long learning for SSFs.

Ironically, therefore, the academic backlash against globalisation of fisheries is using global methods to combat it. As Chuenpagdee (2019, pp. 18–19) notes, "Despite the geographical spread, a large portion of small-scale fisheries is globally connected through mobility and migration, markets and trade, and financial institutions and governance. The 'global' nature of small-scale fisheries calls for concerted research, capacity building, and governance efforts from key fisheries stakeholders and supporting organizations around the world, which is what TBTI aims to offer". The TBTI has established a 'global partnership' to coordinate a world-wide campaign for SSF with five regional networks to facilitate its work in Africa, Asia and Oceana, Europe, Latin America and the Caribbean, and North America. "TBTI research clusters are poised to provide evidence and directions to support small-scale fisheries sustainability worldwide" (Chuenpagdee 2019, p. 23).

This explosion of interest in SSF has triggered five different approaches to its study: poverty reduction; sustainable livelihoods; human rights; well-being; and social struggle. The poverty reduction approach is linked to Béné's argument (2003) that the poverty suffered by the SSF is less associated with biological factors such as over-fishing or economic factors such as open access (the 'old paradigm' of Hardin), than with social factors such as class and ethnic divisions, and political factors such as exclusion of SSF from fisheries management decision-making (the 'new paradigm'). The old paradigm is based on two assumptions: (1) that SSF are poor because they are fishers (who have always over-exploited their fish stocks); and (2) SSF are fishers because they are poor (fishing is a last resort for the poor). These two assumptions lead to the conclusion that SSF is synonymous with poverty.

Béné (2003) refutes this old paradigm and puts forward a new paradigm—that there are many socio-economic and political reasons why SSF are poor but that as a group, they bring great value to a community. Through Béné's (2003) new paradigm, four processes of discrimination against SSF have been identified to explain their poverty: economic exclusion, social marginalisation, class exploitation, and political disempowerment. Economic exclusion entails exclusion from a fishery or potential fishing ground because of entry costs; social marginalisation entails privileged access to a fishery for those with high ethnic, religious or caste status; class exploitation entails economic slavery; and political disempowerment entails being left out of decision-making. Béné (2003, p. 968) concludes that "the different socio-institutional mechanisms that govern the command of fishery resources (essentially social positions and the institutional arrangements controlling the access to, and the use of, these resources) play a more critical role in determining poverty…than purely economic or biological considerations". In other words, the basic cause of poverty in SSF is not their low productivity but their exclusion from economic institutions and public services. Social protection policies, therefore, including social assistance, social insurance, public works, and social justice, will do more to lift SSF out of poverty than would investment in fisheries technology for their economic growth (Béné et al. 2015a). Examples of such social protection policies include scholarships for fishers' children in Mauritius and fuel subsidies/bad weather allowance in the Seychelles. Béné (2009a, p. 912) says SSF are traditionally regarded as poor: "In fisheries, the conventional perception conveyed by the literature is that fisheries (especially small-scale fisheries in less-developed countries) and rural poverty are intimately correlated", but he argues that a more accurate interpretation is that SSF are vulnerable: "fisherfolks are not necessarily the poorest of the poor in monetary terms, but may, instead, be amongst the most vulnerable socio-economic groups, due to their particularly high exposure to certain natural, health-related or economic shocks and disasters". In this respect, SSF in developing countries may be closer to SSF in developed countries, since the latter are very vulnerable to economic shocks (as we shall see), though they are rarely in a condition of abject poverty (although the corona virus Covid 19 may reduce some SFF in the west to temporary poverty).

The second approach to SSF is the sustainable livelihoods approach (SLA). The SLA approach focuses on the way that SSF provides sustainable livelihoods for its fishers and their families and communities. It widens the remit from poverty reduction to a more holistic assessment of the heterogeneous needs of households. As Weeratunge et al. (2014, p. 265) put it, "The 'sustainable livelihoods' concept introduced an intrinsically asset- and capabilities-based framework (combining human, physical, natural, social and financial capitals) to understand how and why people chose…particular livelihood pathways and strategies". Allison is credited with showing how the SLA approach provided a more nuanced understanding of SSF than was provided by the orthodox but simplistic fisheries developmental maxim of making SSF more efficient. In particular, Allison and Ellis (2001) drew attention to the part-time working patterns of many SSF, their diversity of household livelihood strategies, and their complex responses to changes in ecological conditions and fisheries policies. The SLA switched the attention of development from a failed policy of indirect poverty alleviation whereby SSF were encouraged to modernise and sign up to market-based incentives thereby contributing to economic growth which would trickle down benefits to the poor, to a new policy of direct poverty alleviation by improving the living conditions of the poor (Béné et al. 2007). According to this livelihood approach, Hardin's theory that free access to a natural resource will eventually destroy the resource and the livelihoods that depend on it is rejected in favour of Sen's theory of the necessity of the commons. As Jentoft (2011, p. 367) puts it: "Whereas Hardin perceives the freedom of resource users as the problem underlying resource degradation and poverty, Sen regards freedom as an essential part of the solution to the problem". Sen (2000) argues that free access to the sea is not a curse but a blessing for SSF, as it gives them a lifeline to survival. Jentoft and Midré (2011) note that Sen's freedom is not a carte blanche for the SSF to overfish, but a freedom to allow them to choose for themselves how they would regulate the marine commons for long-term sustainability of the fish stocks and thereby their own livelihoods. This is a perception that problems of SSF can only be solved by those who suffer most from them.

SLA theorists contrast the 'wealth-based' approach of industrial fisheries to the 'welfare-based' approach of SSF: "one is geared toward maximization of the rent for the benefit of a few highly capitalistic entrepreneurs, the other is aimed at ensuring that the largest number of poor households can still live from their activities without jeopardizing the ecological sustainability of the resource-base" (Kolding et al. 2014, p. 8). The SLA argues that the harmful way that some SSF treat their fisheries resources (e.g. by overfishing) is not the source, but a symptom, of their poverty, and that the solution to that poverty is not to limit SSF's access to coastal fisheries but to address directly the causes of their poverty, which lie in poor welfare, health and education services (Allison et al. 2011) as well as the damaging impact of industrial fisheries and wider industrialisation. As Kolding et al. (2014, p. 5) put it, "the most productive interventions to promote sustainable resource use and conservation in many communities may lie outside the 'natural resource management system', in domains such as education and health services, or institutions and governance to address the lack of secure access to alternative productive assets".

The third approach to SSF is the human rights approach. An increasing number of researchers are arguing for a human rights approach to SSF (Charles 2011b). Jentoft and Chuenpagdee (2015a) note that SSFs are often poor, marginalised, and therefore vulnerable people, for whom human rights protection is especially needed. The SSF Guidelines (i.e., the Voluntary Guidelines for Securing Sustainable Small-Scale Fisheries in the context of Food Security and Poverty Eradication endorsed by the FAO's Committee of Fisheries in 2014) are founded on the human rights-based approach (Jentoft et al. 2017a). FAO (1995, article 6.18) confers human rights on SSFs in the following statement:

> Recognizing the important contributions of artisanal and small-scale fisheries to employment, income and food security, States should appropriately protect the rights of fishers and fishworkers, particularly those engaged in subsistence, small-scale, and artisanal fisheries,

> to a secure and just livelihood, as well as preferential access, where appropriate, to traditional fishing grounds and resources in the waters under their national jurisdiction.

Charles (2011a) reports that the World Forum of Fisher People (WFFP) and the International Collective in Support of Fishworkers (ICSF) have declared that a human rights-based approach to SSF recognises the inherent rights of fishing communities not only to nutritious food (right of food security), access to fisheries resources (right of sea tenure), clean water, personal safety, decent work, and adequate housing, but also to stakeholder participation in fisheries management decision-making—i.e., empowerment and inclusion (see also Charles 2013). More broadly, the human rights-based approach demands protection for human dignity, cultural identity, non-discrimination, equality (including gender equality), equity, the rule of law, transparency, and accountability. Allison et al. (2012) argue that until and unless fishers' fundamental human rights are secured, management of SSF will fail, whatever the form of governance; and, conversely, that if fishers' fundamental human rights are secured, virtually any form of SSF management will succeed. Willmann et al. (2017b) refer to the influence of the work of Amartya Sen on the human rights-based approach to SSF by forging the conceptual link between human rights and human development discourses. Much of the human rights-based analysis is pitched at developing countries where the grossest violations of human rights such as the right to nutritional food and the right not to be in abject poverty occur, but even in developed countries such as the UK, the language of human rights violations could be used to describe the treatment which some SSF claim they have experienced at the hands of the authorities in being denied access to fisheries resources, equity in the distribution of quota, proportionality in punishment for minor infringements of regulations, and participation in decision-making processes.

However, the human rights approach to SSF has been criticised for two reasons. First, human rights are universal, but fishing rights are specific. Fishing rights invariably have limits (biological, environmental, spatial) attached to them, whereas human rights are unconditional: "human rights are so basic that they defy being bargained or conditioned…When only a specific set of people possesses fishing rights…by definition this negates the claim that fishing rights are akin to human rights" (Song and Soliman 2019, p. 22). Advocates of the human rights-based approach may have to accept some restrictions on the human rights of fishers. Moreover, human rights may conflict with each other. For example, the human right of fishers born in a community to live and fish there may conflict with the human right of migrant fishers fleeing hunger or persecution to live and work in that community. Song and Soliman (2019) ask which human right takes precedence?

The second criticism of the human rights approach is more controversial—that it gives support to neo-liberal policies based on property rights. For example, Ruddle and Davis (2013, p. 87, 89, 92) claim that the human rights approach "facilitates a neo-liberal agenda…employing human rights rhetoric to discredit customary practices and to separate the individual from her/his social context and relationships… human rights have been co-opted as a central ideological cornerstone in furtherance of the neo-liberal development agenda". According to Ruddle and Davis, the human rights discourse justifies the neo-liberal doctrine of individual transferable quotas (ITQs) and the privatisation of the commons, and it fails to protect the fishing rights of indigenous people. Ruddle and Davis (2013, p. 91) argue that a genuine rights discourse identifies SSF rights not in an abstract theory of universal individual rights, which lend support to property rights like ITQs, but in the concrete context of historical, cultural, communal, entitlements: i.e., traditional customary rights:

> the 'rights' in SSF settings are best approached as based on arising from the history, processes and dynamics of cultural expressions and social relationships represented in SSF peoples' identities, understandings, practices, and ways of living. These are usually associated with specific geographical locations. Consequently, these 'rights' are collective and, as such, inalienable, i.e. cannot be bought or sold by individuals engaging with and benefiting from such 'rights'. They are invested in, arise from, and are expressed through, group cultural and social relationships. In this sense, they are rights arising from, and embedded in, culture, and not explicable as commodifiable and transactable property rights.

At the heart of Ruddle and Davis' argument is a rejection of the concept of private property in catch shares, especially enshrined in the form of individual transferable quotas (ITQs). ITQs are an iconic symbol of the neo-liberal approach to fisheries management, introduced to reduce overfishing by giving quota holders a vested interest in protecting fish stocks as their own assets. Isaacs (2015) argues that one of the major threats to justice for SSFs is the introduction of ITQs because it opens the door to the concentration of quota in the hands of an increasingly narrow elite of wealthy industrial fishers and processors (see also Jentoft and Chuenpagdee 2018; Bodwitch 2017; Lalancette 2017; Allison et al. 2012). NUTFA [New Under Ten Fishermen's Association] (2019a, b) states that "ITQs make inshore fishermen increasingly reliant on the 'investor class' that has little or no interest in anything other than a return on capital". Pinkerton (2019, p. 242) states that "ITQs turned access rights to fish into tradable commodities, based on the idea that the market was a better instrument than government for allocating access rights". The resulting impact on SSF as a consequence of governments taking this forward has been severe. "Allocating fishing rights to individual small-scale fishers in South Africa resulted in the elite of the community obtaining fishing rights while the majority of bona fide fishers were overlooked" (Isaacs 2006, p. 52). Eide et al. (2011, p. 22) asks: "is it not better to prioritize the livelihoods of the millions depending on small-scale fishing rather than maximizing wealth creation for a relatively small and already privileged group of better-off fishers?" Jacobsen (2013, p. 3) claims that "Since the 1980'ies, market-based fisheries management systems based on ITQs have gradually replaced traditional state-based command and control regimes in developed-nations fisheries across Oceania, North America and Europe".

Høst (2015) states that ITQs have resulted in the serious decline of the Danish SSF. Chambers and Carothers (2017) claim that the ITQ system introduced in Iceland thirty years ago has harmed SSF communities, created inequality between quota holders and non-quota holders, and increased the rate of illegal discards. They argue that privatisation is based on the neoliberal assumption that fishers are self-interested profit-maximisers, which is not the case for SSF who are committed to communitarian and environmental goals rather than individual enrichment. Davis

(2015) reports how in Newfoundland during the cod moratorium, SSF managed to fight off a proposed ITQ system, but since then, increasing corporate control is threatening the independence of SSF. Edwards and Edwards (2017) describe how the introduction of an ITQ regime in British Columbia in 1990 has resulted in a concentration of quota in the hands of investors with no connection to fishing who lease quota to SSF at exorbitant cost (see also Pinkerton and Edwards 2009). Pinkerton (2015, p. 413) refers to "The type of extreme inequality in the fishing fleet which came to exist in neo-liberalized North American fisheries" under which "Small-scale fishermen who are lessees of ITQs may be called 'serfs' and quota owners may be called 'fish lords'". She claims that "resource economists who support ITQs first and foremost for the efficiency they are purported to deliver implicitly acknowledge that they have not been concerned with distributional issues (equity), and that equity has been sacrificed to efficiency" (see also Edwards and Pinkerton 2020). Jentoft and Johnsen (2015) report moves to slim down the SSF in Norway to a fraction of its former size by means of a transferable quota scheme. Kolding et al. (2014, pp. 7–8) say "the restructuring of the Norwegian fisheries (that supported more than one hundred thousand small-scale operators until WWII) into the current wealth-generating, highly capitalized and efficient fishery has required a 90% reduction in the number of fishers". Jacobsen (2013) reports similar effects on SSF in Greenland. Song et al. (2018) describe how in New Zealand, the introduction of ITQ in 1986 marginalised SSF, including Maori, who were excluded from their traditional fishing areas. In a settlement reached in 1992, Maori were compensated for this exclusion by being allocated a collective quota, but this locked them into the ITQ system which has left many Maori SSF still marginalised. Stewart et al. (2006) record that of the 3000 fishers who exited from the New Zealand after 1986, most of them were SSF. Eide et al. (2011, p. 22) write that "Individual property right systems may improve profitability, but will most likely be devastating to the small-scale sector". LIFE (2016, p. 1, 16) claims that:

> Where they have been imposed around the world, the experiences have been consistent: ITQs have led to loss of employment, concentration of ownership, and increased social and economic costs for SSF...many of the purported benefits of ITQs are hypothetical, false, ideologically motivated, or exaggerated...ITQs are a system based on an economic logic that does not take into account wider impacts and different forms of value. They make access to fisheries more difficult for small-scale fishers, prevent access for new fishers, result in economic consolidation and drive inequality, and negatively impact small coastal communities.

It is true that some other researchers argue that ITQs do not always harm SSF. For example, Johnson and Pálsson (2015) claim that ITQs have been beneficial for SSF in the Lake Winnipeg fishery. It is also true that steps can be taken to shield SSF from the harmful effects of ITQ. For example, Soliman (2015) reports that in Alaska the ITQ system has been modified to protect SSF, and Edwards and Edwards (2017) explain how in British Columbia, quota banks have been created by fishers to mitigate the effects of ITQs on SSF. Foley et al. (2015, pp. 391, 390) found evidence in a Canadian shrimp fishery that "enclosure, quota renting, and commodification—concepts usually associated with exclusionary consequences of neoliberalism—can be regulated to empower communities and to embed fisheries resources and development benefits in coastal communities…principles of ethical allocation and distribution of resources are able to persist through an era of neo-liberalism". Langdon (2015) suggests ways in which SSF in southern Alaskan coastal villages could work around ITQs to exploit unharvested salmon. Some writers argue that the human rights approach, far from giving succour to neo-liberal notions of private property rights such as ITQs, can be used to insulate SSF from the harm that ITQs may otherwise inflict on SSF. Willmann et al. (2017a) claim that the human-rights-based approach actually protects the vulnerable from the pernicious inequality generated by privatisation. Indeed, according to Song and Soliman (2019, p. 20), the human-rights discourse is explicitly designed to counter the property-rights discourse: the human rights approach has arisen

> out of the rejection of the so-called 'rights-based' approach, in which fishing rights are privatized and individuated to become an ownership-based, tradable commodity (e.g., Individual Transferable Quotas, ITQs) and therefore serving as a neoliberal basis for few, select right-holders' wealth accumulation. The framing of seeing fishing rights as human rights is…to remind us that the rights of small-scale fishers to access and harvest fishery resources are, in many cases, a fundamental and indivisible part of their culture, survival, and wellbeing, which ought not be capitalized, sold, and otherwise removed from their social-historical contexts…understanding fishing rights as integral to fishers' human rights would provide a strong rhetoric to fend off the socio-economic exclusion and marginalization undergirded by the prevailing trend of 'propertization' of fishing rights.

One way of using the human-rights approach to counter ITQs would be to invest customary rights of sea tenure with human rights credentials. Willmann et al. (2017a) claim that the human-rights-based approach supports the idea of community rights such as collective rights of sea tenure, which is very beneficial to many indigenous SSFs across the world. Davis and Wagner (2006) suggest a historic use rights approach to SSF: referring to coastal fisheries in Canada, they point out that indigenous people have been granted property rights based on their historically continuous use of fisheries resources in specific localities. Similarly, they argue, non-indigenous SSF should have a right to fish if they satisfy certain conditions of historic usage. Pinkerton (2015) suggests a 'moral economy' approach to SSF which she connects to the human rights discourse. The concept of moral economy refers to traditional norms and values which have underpinned practices in SSF for generations. For example, the catch share system whereby the profits from catches are shared out amongst all the crew in an equitable fashion is an embodiment of a principle of fairness which treats everyone with dignity and equal respect and ensures their welfare and well-being. The moral economy is thus a "moral code which was not trumped by profit maximization" (Pinkerton 2015, p. 411).

The fourth approach to SSF is the well-being approach, which provides an overarching framework for establishing what is important for individuals and communities:

> 'Wellbeing' constitutes a broad-based *outcome*, including material goals, such as economic yield, food supplies and employment, as well as non-material aspects – for example, safe, decent and non-discriminatory work conditions in fisheries…or preservation of ecological

> values of marine and coastal ecosystems…As an outcome, wellbeing provides a broader conception of social benefit than one that is typically captured in current fisheries frameworks (Weeratunge et al. 2014, p. 255).

In the past, well-being was characterised as 'well-having'—i.e. not being in poverty; having a livelihood; or possessing rights—but now well-being is characterised as 'well-living'—i.e. living a life well-lived or living the kind of life one values—and this requires subjective judgement about the meaningfulness of one's life: "This reflects a well-articulated argument to go beyond a traditional poverty or deprivation-centred lens in looking at how people envision living their lives, choose to live these lives and what they value in their lives" (Weeratunge et al. 2014, p. 270). The well-being approach compels us to take into account that SSF regard fishing not merely as a means of ensuring livelihoods, but as an inherently rewarding activity in its own right—as a fulfilling and meaningful way of spending one's working life. Weeratunge et al. (2014, p. 269) report that McGoodwin argued that "although 'economic rationality' might not explain why fishers tenaciously adhere to an occupation with diminishing returns, 'existential rationality' perhaps does". Weeratunge et al. (2014, p. 272) claim that the well-being approach can be applied to fisheries management policies, not least in cautioning managers about removing SSF from fishing: "a well-being lens may be useful in assessing the potential success or failure of policies aiming to move fishing households out of fisheries, whether as a response to depletion of stocks or as a policy preference for industrial over small-scale fisheries, by highlighting facets such as job satisfaction, identity and fishing as a way of life". However, they acknowledge the difficulty of obtaining data on, and measuring, well-being (Weeratunge et al. 2014, p. 272).

The fifth approach to SSF is to see fisheries as a site of social struggle, which is the most recent approach. Bavinck et al. (2018) enunciate this approach, arguing that fisheries researchers have traditionally concentrated on the issues of the sustainability of fisheries (environmental concern) and the management of fisheries (governmental structure), but have neglected the issue of social struggle (distributional justice). This is regrettable, because social struggles over fisheries are increasing across the world, and they have a direct impact on fish stock health: "addressing distributional justice concerns may be a precondition for achieving sustainable human-nature relations" (Bavinck et al. 2018, p. 46). Bavinck et al. argue that capture fisheries are becoming a deepening zero-sum game whereby the gains of one fleet are achieved at the expense of the losses suffered by another fleet because the fishing opportunities are narrowing as competition for fishing space increases. In particular, there is a struggle between an increasingly dominant industrial fleet and an increasingly weak artisanal fleet: "Some of the most momentous and widespread social struggles that have occurred in the fisheries relate to the industrialization process as it has unfolded in fisheries since the middle of the 19th century" (Bavinck et al. 2018, p. 48). The authors urge social scientists to study these social struggles because they "alert us to tensions and inequalities within society" which "can be the forerunner of violent conflict and a reduced governability of fisheries" (Bavinck et al. 2018, p. 50).

Each of these five approaches is teleological in that they all presuppose an end or purpose that SSF serves. For example, the poverty reduction approach sees SSF as a means to reduce penury; the sustainable livelihoods approach sees SSF as a means to achieve secure employment; the human rights approach sees SSF as a means to treat people with basic respect; the well-being approach sees SSF as conferring contentment; and the social struggle approach sees SSF as fighting against injustice. While the present book takes account of all five approaches to SSF, its focus is less on an end to be achieved, than on the strategies used by SSF to achieve those ends. So the book adopts a sixth approach—resilience theory. We explain in detail in the next chapter what resilience theory means and its value in interpreting SSF, as well as some of the criticism levelled against it. Suffice it here to say that the book investigates the case for the sustainability of SSF by means of a detailed study of the resilience of the SSF in England, known as the 10 m and under fleet or the inshore fleet, which has been subject to many of the pressures common to SSF elsewhere in the world. The purpose of the book is to examine these pressures and assess what strategies these English inshore fishers have taken to withstand them. We found three strategies adopted by most fishers—passive resilience; adaptive resilience; and transformative resilience. The book analyses these three strategies and assesses their viability, addressing the question of whether they will enable the inshore fleet to survive competition from industrial fisheries, impacts from marine industries such as offshore windfarms, and spatial restrictions imposed for environmental conservation purposes. Chapter 2 sets out the interpretive framework of the book in terms of resilience theory, and explains how primary data were obtained. Chapter 3 outlines the history of the English inshore fleet to provide insight into its current situation. Chapter 4 analyses the challenges faced by English inshore fishers. Chapter 5 explains the three ways in which English inshore fishers have responded to these challenges. Chapter 6 discusses the most important issues raised by the findings in Chaps. 5 and 6. Chapter 7 concludes the book with a summary of its findings and some recommendations to policy makers and managers on ways to help SSF to survive in England. But first, we discuss the value of SSF globally; the challenges facing SSF across the world; and the coping strategies they have used to meet those challenges.

1.5 The Value of SSF Globally

Fishing can be counted amongst the world's oldest livelihood options (Thompson et al. 1983), currently employing roughly 50 million people globally (Salmi 2015). It is estimated that more than 95% of those employed in capture fisheries are small-scale fishers, with most residing in developing countries (FAO 2007; Urquhart et al. 2013; Eide et al. 2011; Scholtens 2016). Not included in these estimates, however, are hundreds of millions of people for whom fishing is not a full-time occupation but represents one component of a multi-activity livelihood strategy (FAO 2005). Symes and Phillipson (2009) explain that SSFs also provide many forms of informal labour without which many small family businesses engaged in the harvesting side of fisheries together with the processing, marketing and other supportive structures would not survive (see also Jentoft and Chuenpagdee 2015a; FAO 2005; Morgan 2013). According to Basurto et al. (2017, p. ix), SSF are "far and away the ocean's largest employer (greater than oil and gas, shipping, tourism etc.)", and the SSF sector provides more jobs per unit of fish landed than does the industrial sector (Percy 2014).

The literature suggests that SSFs contribute between 15% and 30% of total world fisheries production (Chuenpagdee and Jentoft 2011, 2015) and approximately 46% of all capture fish (Scholtens 2016). Almost all fish from SSF is used for direct human consumption (Zeller and Pauly 2005; Pauly 2006), in contrast to industrial fisheries where a substantial percentage of its catch is used for animal feed and other products rather than for direct human consumption (FAO 2005). SSF operations also make vital (if inadequately recorded) contributions to the food security of many millions of people in fishing communities (Lloret et al. 2018; Akintola and Fakoya 2017; Andrew et al. 2007; Nayak and Berkes 2010; Berkes et al. 2001; Jentoft 2011; UNDP 2005; World Bank/FAO/WorldFish 2010). Catches from SSF often bring to the table the most important source of protein available for poor people, contributing more than 50% of animal protein intake in many developing countries (Basurto et al. 2017). Therefore, the small-scale sector plays a central role in feeding the poor and vulnerable (Jentoft and Eide 2011; Salas et al. 2011). As Willmann (2015, p. ix) put it, "the importance of small-scale fisheries can hardly be overestimated for food security, nutrition, livelihoods, rural development, and poverty reduction" (see also Pinkerton 2019). According to Béné (2006), the main contribution of most SSF is welfare provision rather than financial profit and it also improves the distribution of wealth among the various social groupings engaging women, children, and people with few other options to sustain themselves (see also Béné et al. 2007).

Eide et al. (2011) say SSF serves as a 'safety valve' or 'safety net' for the rural poor, helping to stem emigration by providing employment opportunities for women and young people (see also Kolding and van Zwieten 2011). SSF provide a buffer against destitution when families face emergency situations such as when the family head loses their job or crops fail or civil war or natural disasters force people to flee their communities (FAO 2005). So instead of seeing SSF as the cause of over-fishing and therefore poverty, we should see SSF as a means of overcoming poverty (Eide et al. 2011) and facilitating self-reliance in a situation where a safety net is not provided by their government.

Whilst it is true that SSFs exert significant pressure on fisheries resources to the point of overfishing (Carvalho et al. 2011), according to Béné (2006) their negative impact on global stocks is no greater than that exerted by industrial fisheries. Moreover, the environmental impact of SSF is far less than that of industrial fisheries (Lloret et al. 2018; Kolding et al. 2014; Chuenpagdee and Jentoft 2015) as they use less damaging gear, target many species rather than chasing a single species, and work with a lower carbon cost per kilo of fish landed (Percy 2014). Importantly, SSFs are often active in stewardship initiatives and conservation efforts to sustain their immediate surroundings (Chuenpagdee and Juntarashote 2011; Symes 2002).

SSF often enforce self-imposed rules and have higher levels of marine stewardship than larger commercial vessels (Gray et al. 2011). SSFs claim they work sustainably within the area they fish and are repositories of valuable traditional knowledge of fish stocks and marine ecology (FAO 2005; Johannes 1982; Symes and Phillipson 2009).

Moreover, the oft-quoted description of fishing as a 'way of life' (Aryeetey 2002; Urquhart et al. 2013; Morgan 2013; Jacob et al. 2001; Nuttall 2000; Trimble and Johnson 2013), rather than merely a job, remains applicable to many SSFs (Morgan 2013) where values, knowledge and traditions are established and passed on between fishers (Brookfield et al. 2005; Van Ginkel 2001). In addition to providing income, SSF is held to be an interesting, challenging, and independent vocation (Kraan 2011; Nunoo et al. 2006) from which fishers gain a strong sense of pride, social identity and solidarity (Apostle et al. 1985; Gatewood and McCay 1988; Onyango 2011; Pollnac 1979). Urquhart and Acott's (2013) study indicated that "they were proud of their identity as fishermen". A fisher from Pollnac and Poggie (2006)'s study said it was not the money that was important to him, but the job. He defined himself by his job, and if he could not fish, he would not be himself—he would lose his identity. This can be seen in SSF's deep-seated desire to fish even when it is no longer economically viable (Van Ginkel 2001). Pollnac and Poggie (2006) suggest that this is tied to psychological characteristics (adventurous, active, most aggressive and courageous risk takers) which sit at the core of the personalities of SSFs.

Furthermore, there is evidence to suggest that inshore fisheries have a strong influence on the social capital of local coastal communities (Symes 2009). Crews tend to be chosen from close family or community ties, and this is important for generating enduring egalitarian and reciprocal community relationships (Symes 2001). SSF values manifest themselves through contributions to the social and cultural fabric of the fishers' communities (Morgan 2013; Onyango 2011). Allen (2013) claims that SSF are the foundation of many Canadian coastal communities. Brookfield et al. (2005) say the inshore fleet is the medium by which community values and traditions are maintained and passed on. Narratives of SSF have been interwoven in the fabric of many communities, enhancing their cultural heritage, expressed through popular myths, folktales, and local history (McGoodwin 1990; Gudeman 2001; Urquhart and Acott 2013; Aryeetey 2002; Jentoft and Eide 2011). Acott and Urquhart (2014) describe the cultural value of SSF as giving 'a sense of place' to residents and visitors in coastal communities (see also Reed et al. 2013; Urquhart and Acott 2014; and Urquhart and Acott 2013). Béné (2006) refers to the sense of self-esteem generated by this cultural identity. Vaughan and Vitousek (2013) explain the immense cultural value of the non-material benefits that subsistence SSF provides to Pacific communities through the non-commercial distribution or sharing of fish.

Nevertheless, despite all the above positive features of SSF, there are some negative features. Johnson (2006, p. 753, 747, 754) warns of the "pitfalls of romanticism", and says we must not assume that SSFs "inherently embody social justice and ecological sustainability", because they can be "highly exploitative" and "ecologically destructive". For example, he notes that "small-scale fishers in the

Philippines and Indonesia have exacted a terrible toll on the countries' reefs using dynamite and cyanide fishing". Kolding et al. (2014, p. 3) say SSF is "associated with 'Malthusian overfishing'...indiscriminate and unselective fishing methods, illegal and destructive gears...overcapacity and resource depletion". Lloret et al. (2018, pp. 176) write that "the use of fishing gears that actively select certain species, sizes and sexes, the deployment of fishing gears on certain fragile habitats, the loss of fishing gears, and the use of non-native species as bait, are examples of how SSFs can threaten the sustainability of vulnerable coastal species and habitats".

1.6 The Challenges Facing SSF Globally

Small-scale fisheries across the world are widely considered to be in a state of crisis (Campling and Havice 2014; Berkes et al. 2001; Salas et al. 2007; Jentoft and Eide 2011; Urquhart et al. 2013) including in Europe (Lloret et al. 2018). This crisis has major impacts, resulting in the deterioration of SSF communities and many problems of equity as well as environmental damage (Bavinck 2011; Nayak and Berkes 2010). Much of the literature on the SSF crisis is focused on developing countries (Béné 2009a, b; Chambers 1989; Khan 1998; World Bank 2000; Prowse 2003; Clay and Olson 2008) and their high levels of exposure to economic shocks and natural disasters (Allison et al. 2006, 2009). Several challenges facing SSF in developing countries may not be faced by SSF in developed countries—including abject poverty, acute food insecurity, increased migration to coastal areas, absence of fisheries data, and lack of governance. But many of the challenges they face are echoed in developed countries, and the two sets of SSF may be equally vulnerable to those challenges. Such challenges include lack of lending and micro-credit facilities (Salas et al. 2011; Jacquet and Pauly 2008); inability to respond to technical change (Thorpe et al. 2007); ecological fragility of the fishery (Cinner et al. 2009); insecure access to markets, fisheries, and equipment (Thorpe et al. 2007); overdependence on fishing (Thorpe et al. 2007); lack of alternative livelihood options (FAO 2005; Salas et al. 2011; Cinner et al. 2009); overexploitation and depletion of resources (Chuenpagdee and Jentoft 2015; Jentoft and Eide 2011); conflicts over resources and coastal space especially with large commercial fishing operations (Andrew et al. 2007; Jentoft and Eide 2011; Chuenpagdee and Jentoft 2015); cultural and class tensions (LiPuma and Meltzoff 1997; Meltzhoff and Schull 1999; Allen and Gough 2006; Kitner 2006; Blount 2007; Masozera et al. 2007; Aizenman 2007); globalisation of markets (Jentoft and Eide 2011; Chuenpagdee and Jentoft 2015); geographic remoteness of many communities (Thorpe et al. 2007); fluctuating fuel prices (Jones et al. 2010; Abernethy et al. 2010); climate change (Andrew et al. 2007; Jentoft and Eide 2011); the high risk nature of SSF fishing activities (Thorpe et al. 2007); pollution and environmental degradation (FAO 2005); inappropriate fisheries management policies (Jones et al. 2010; FAO 2003; Crosoer et al. 2006); weak stakeholder representation and participation (Chuenpagdee and Jentoft 2015); lack of social capital (Kosamu 2015); corruption (Chuenpagdee and Jentoft 2011;

Islam 2011); gender inequality (FAO 2015); and political marginalisation (Thorpe et al. 2007).

From this list, the seven most serious challenges to SSF identified in the literature are discussed below: the first four are externally caused challenges; the last three are self-inflicted. The four external challenges are ocean grabbing; political exclusion; lack of diversification; and weather hazards. Ocean grabbing (Bavinck et al. 2017; Bennett et al. 2015) means SSF are ousted from the commons on which they depend (Sen 2000; Beck and Nesmith 2001; Nayak and Berkes 2010). This can be attributed to the re-allocation of coastal resources to other uses such as recreation and tourism (Chuenpagdee and Jentoft 2015) industrial fisheries, aquaculture and other marine-based industries, and the spread of protected areas with precautionary management arrangements. As coastal development has mushroomed, SSF communities have increasingly suffered economic and social disruption (Apostle et al. 1998; McGoodwin 1990; Johnson and Orbach 1990; Acheson 2003; Clay and Olson 2008). This is aggravated as traditional SSF communities may not have legally recognised property rights to an area despite using the area for many generations. As Bennett et al. (2015) note, this is a serious problem for SSF, threatening human security, economic livelihoods, and socio-ecological well-being. For example, the Mediterranean has favoured tourism activities to the detriment of artisanal fisheries, and the marine-based recreational activities associated with a rise in tourism (e.g., boating, diving and angling) have driven fishers away from their fishing grounds in peak seasons (see also Gómez et al. 2006). The development of industrialised fisheries alongside SSF has frequently resulted in tension and strife (Berkes et al. 2001; FAO 1999). The larger vessels may push aside small-scale vessels as they do not respect traditional tenure or restrictions set by governments (Amarasinghe and Bavinck 2011; Jentoft et al. 2010; Tvedten and Hersoug 1992; Chuenpagdee and Jentoft 2011) or tow or damage set gear which is expensive for SSF to replace. Large-scale vessels may wipe out commercially important coastal fish stocks, sometimes by illegal fishing activities (Jones et al. 2010). "Illegal fishing is one of the biggest challenges for SSF and a major threat to their viability" (TBTI 2017, p. 12). Many writers raise the issue of illegal, unreported and unregulated (IUU) industrial vessels especially in developing countries' waters (Doumbouya et al. 2017; Intchama et al. 2018). IUU vessels compete directly with the SSF sector by targeting similar species, and entering areas that are reserved to SSF operators, and this has a major impact on fish stocks (Belhabib et al. 2015). Ocean grabbing also includes the activities of major marine-based industries. For instance, in the Gulf of Mexico, a zone of exclusion for all but the oil industry was established in 2003 which restricted SSFs' access to their fishing grounds and forced them to travel further out to sea, making their fishing operations more demanding and dangerous (Quist and Nygren 2015). Aquaculture is another major competitor for water area usage at the expense of the traditional fishing grounds of local SSF (Nguyen and Flaaten 2011; Ben Tre Dard 2009; Nayak and Berkes 2010). Also, the transmission of diseases from aquaculture to wild fish has had a significant impact on the stability of SSF fishers' incomes (Ben Tre Dard 2009). The establishment of marine protected areas (MPAs), which is regarded by many researchers as the most effective tool for conserving biodiversity (Dudley 2008), is seen by many SSF as another form of ocean grabbing, because they have experienced extensive restrictions on their access to landing places and work space by such spatially-based protection efforts (Weyl and Weyl 2001; Chuenpagdee and Jentoft 2015). SSFs argue that governments which establish MPAs often do not appreciate the importance of small-scale fishing, placing the protection of marine organisms above the well-being of fisheries-dependent coastal communities (Mascia et al. 2010; West et al. 2006; Faraco et al. 2016; Thomas et al. 2014; Leis et al. 2019; Pinkerton and Davis 2015).

As the above section implies, the second external challenge is that SSF have often been poorly governed by fisheries management bodies (Andrew et al. 2007). For example, SSF are under-represented in bodies governing fisheries whose decisions therefore invariably reflect the perspectives of the industrial sector and ignore the needs of SSF (Béné et al. 2015b). SSF are very diverse in terms of participants, resources, gears, geographical spread, scale, and connectivity to other livelihoods (Chuenpagdee et al. 2005; Berkes et al. 2001; Berkes 2003), yet managements tend to adopt a narrow and homogenous perspective with a restricted fisheries governance vision (Chuenpagdee and Jentoft 2015). Degnbol et al. (2006, p. 534) refer to "disciplinary boundaries [which] narrow the perspectives of fisheries management, creating tunnel vision and standardized technical fixes to complex and diverse management problems". For the most part, fisheries policies and governance are directed at the daily tasks that technical and routine solutions may be able to handle (Chuenpagdee and Jentoft 2015; Kooiman 2003). Policies often lack social and cultural objectives (Urquhart and Acott 2013; Symes 2000), and Symes et al. (2015) claim that these restrictions have significantly reduced the flexibility of fishing activity that would enable fishers to adapt to changing ecological conditions. Part of the reason why SSF have been treated inappropriately has been attributed to their weak political status, meaning that they do not attract much political support (Allison and Ellis 2001; Lam and Pauly 2010; Faraco et al. 2016), and therefore face indifference and neglect by their governments (Béné 2003). Many accounts of SSF (Isaacs 2006; Mathew 2003; Andrew et al. 2007) reveal that SSF are commonly under the control of a hierarchical governing system in which policy makers often ignore the needs of small scale fleets (Berkes et al. 2001; Chuenpagdee and Jentoft 2015), favouring instead the capitalist interests of more powerful large-scale industrial fisheries (Prosperi et al. 2019; Crosoer et al. 2006). Rarely do SSF have a voice in forums where their fate is decided (Chuenpagdee and Jentoft 2015).

Lack of alternative or supplementary employment opportunities for SSFs is the third external challenge (Aryeetey 2002; Allison et al. 2011). One reason for lack of movement into non-fishing jobs is inadequate transport infrastructure preventing commuting (Allison et al. 2011). This means that for some SSF fishers it is not only difficult to obtain jobs elsewhere, but it is hard for them to develop their fishing jobs, such as direct marketing—i.e., sell their catch in distant markets which fetch higher prices—so instead they sell the fish at coastal landing sites for lower prices (Islam 2011; Andrew et al. 2007). Other reasons for a lack of alternative employment include little external investment (Nguyen and Flaaten 2011); low education levels (Aryeetey 2002; Horemans 1998); limited transferability of fishery skills; and unavailability of training programmes in other skills (Islam 2011). An inability to gain credit through banks is another contributor to SSF vulnerability (Mills et al. 2011a; Jacquet and Pauly 2008), because without financial capital, SSFs cannot invest in innovative technologies or working capital (Aryeetey 2002). Fishers' access to banks is very limited due to lack of collateral assets like landed property, so they are dependent on informal credit mechanisms (Amarasinghe and Bavinck 2011) such as the dadon system operating in Bangladesh (Islam 2011). These informal arrangements are however, often blamed for exploiting SSF fishers because they bind them to money lenders (Habib 2001) who charge very high interest rates and lock SSFs into a permanent state of indebtedness (Amarasinghe et al. 2005).

In addition to anthropogenic challenges, SSFs and their communities live under conditions that make them vulnerable to big weather events such as cyclones and hurricanes (Chuenpagdee and Jentoft 2011). This is the fourth external challenge. In countries like Bangladesh and Nicaragua, Guatemala, and Mexico, these are common occurrences (Islam 2011; González 2011; Andrade and Midré 2011; Salas et al. 2011). Natural disasters expose SSFs to greater life-threatening dangers than fishers operating in bigger and safer vessels (Ahmed and Neelormi 2008). When disasters strike, limited finances make it difficult for SSFs to restore homes and infrastructure, and the loss of fishing gears, boats, livestock, and other household assets can wipe out entire livelihoods (Islam 2011). Climate change contributes an additional hazard to local livelihoods because it brings increasingly challenging weather conditions making fishing a riskier occupation (Hall 2011; Faraco et al. 2016) with reductions of fish stocks (Cheung et al. 2010); changes in spatial and/or temporal distribution in fish stock (Morzaria-Luna et al. 2014); decreases in the value of landings; and high adaptation costs (Pörtner et al. 2014). According to researchers, the impacts of climate change will be felt predominantly by artisanal or small-scale fishers mainly in developing countries (Badjeck et al. 2010). Of course, all fishers have had to adapt to the vagaries of the weather (Faraco et al. 2016; Glantz and Thompson 1981; Cole 1996; Gordon and Munro 1996; Lauck et al. 1998; Allison et al. 2009; Rothschild et al. 2005) and meteorological uncertainty has always been inherent in fisheries (Miller and Fluharty 1992). However, current and expected rates of change in weather patterns are unprecedented (Faraco et al. 2016; MacKenzie and Schiedek 2007; Dulvy et al. 2008), and SSFs are less able than industrial fishers to adapt to climate change because their limited spatial scale of activities restricts their adaptive capacity (Morton 2007; Ahmed and Neelormi 2008).

There are three self-inflicted challenges contributing to the vulnerability of the SSF: over-fishing, poor self-organisation, and gender inequality. Although large-scale vessels are responsible for most over-fishing, some SSFs work in a Malthusian fashion, running down fish stocks, thereby contributing to the destruction of the resource from which they draw their livelihoods (Lloret et al. 2018; Chuenpagdee and Jentoft 2011; Korda et al. 2008; Islam 2011; Pomeroy 2012; Andrade and Midré 2011). Some SSFs use unsustainable gear, including beach seine nets with small mesh sizes (Nguyen and Flaaten 2011; Hosch 2002; Kraan 2011), whilst others are known to poach protected species or target stocks during sensitive breeding or nesting periods. Over-fishing in developing countries can also be attributed to the virtually open access nature of the industry and the expansionary policy measures of the past which encouraged more people to enter the SSF fishing sector (Aryeetey 2002). The effects of global overfishing are serious, with many stocks already collapsed and others on the way to becoming economically non-exploitable (FAO 2008; Béné 2003; Béné et al. 2004; FAO 2005; Clark 2006; Clay and Olson 2008). Over-fishing threatens not only the future of fisheries, but also the livelihoods of coastal communities (Faraco et al. 2016; Urquhart et al. 2013). Over-fishing means that fishers have to travel farther from shore for sizable fish thereby putting their lives at risk, and they may be forced to migrate away from their communities (Jones et al. 2010). Also there is increasing conflict in the fishery as fishers compete for fewer fish (Faraco et al. 2016; Metzner and Ward 2002; Ward et al. 2004). Pomeroy (2012) argues that dealing with SSF overfishing is much more difficult than dealing with overfishing by large-scale vessels, for three reasons. First, data on SSF numbers, catches, and gear are much poorer than data on industrial vessels (Mills et al. 2011b). Second, instruments for management such as quota controls and modes of enforcement such as patrol boats are not as practicable for SSF as for industrial fleets because of the much greater number of SSF vessels. Third, the impact of cutbacks on the livelihoods of SSF families is much more damaging than the impact on the crew of industrial vessels because most SSFs are unable to move to a different area and have fewer alternative employment options.

Whilst not applicable to all SSF groups, the second self-inflicted challenge is a lack of self-organisation in the SSF sector. SSF associations often lack cohesion and strategies with clear goals, and this weakens their effectiveness in representing the sector in negotiations with management (Allison and Ellis 2001). This deficiency is due partly to the individualistic and independent-mindedness of SSFs and their unwillingness to engage, partly to high subscriptions charged for joining cooperative organisations, and partly due to a lack of social capital (Paladines 2015; Jentoft 2000). On the last factor, in a study of 17 cases of SSF in developing countries, Kosamu (2015, p. 365) concluded that "the sustainability of small-scale fisheries depended solely on the strength of collective social capital of the local communities at the resource scale". Without social capital, the relationship between SSF communities and government is frequently characterised by hostility (Chuenpagdee and Jentoft 2015).

The third self-inflicted challenge to SSF globally is gender inequality. Authors note many factors which prevent women from performing direct fisheries roles, and confine them to the lowest end of the fishing value chain. One factor is the culture of exclusion, as these fishing roles are seen as non-traditional occupational pursuits for women (Fröcklin et al. 2013; Yodanis 2000). Women report difficulty finding ships to work in; the lack of female-friendly accommodation and facilities on-board; and incidents of bullying, harassment and cultural taboos created by the male-dominant ship culture (Zhao et al. 2013). Unsociable working hours is another factor for women to grapple with because of their childcare responsibilities, and Kaplan (1988) notes that in order to go to sea, women have to work harder than the men to prove their capability and earn their respect as many men deem the profession to be unsuitable for women. Also, women are under-represented in SSF-related research (Kleiber et al. 2014; Kaplan 1988). Data on fishing are largely gender-disaggregated (Fröcklin et al. 2013): few data sets contain figures of female workers in the fishing industry (Kleiber et al. 2014) and official figures do not identify female-held positions in marketing, accounting, management, or fishing association leadership (Zhao et al. 2013). This means that women's ecological knowledge remains an untapped resource (Kleiber et al. 2014). With the exception of a few feminist authors (Yodanis 2000; Nadel-Klein 2000; Fröcklin et al. 2013; Chuenpagdee et al. 2006; Zhao et al. 2013; Sharma 2004), the literature associates small-scale fishing with male-only activity. Furthermore, even when women's roles are discussed, the focus is usually on how they help men to be resilient (Kaplan 1988; Britton 2012b), and this shows their contributions are given lower cultural status or economic value and are viewed as inferior to the men's work (Kleiber et al. 2014; Yodanis 2000). Nadel-Klein (2000) found that women's stories are often omitted from formal records: for example, the Fisheries Museum of the Atlantic allocates no exhibit space for women's activities in the industry as they are seen as home makers rather than fish workers. Unsurprisingly, therefore, women's paid and unpaid contributions to SSF are undervalued and almost invisible in contemporary policies, governance arrangements, management plans and fisheries statistics (Fröcklin et al. 2013; Kleiber et al. 2014; Britton 2012b). Furthermore, they do not participate in decision making processes (Zhao et al. 2013), which limits their ability to influence decisions which affect their lives, and so decisions that are made rarely represent women's interests. Ignoring women's roles (secondary and primary) in SSF marginalises them, and when women are marginalised, SSF communities are also marginalised. Gender neutral terms (such as 'fisher') have been adopted throughout this book as both women and men are involved in all aspects of the SSF fishery and were included in the research scope. However, we recognise that some women dislike the word 'fisher' because of its negative connotations of something fishy, and prefer terms like 'fisher-folk'. Tellingly, one female FA lead pointed out that women who worked in the fishing industry had never been asked what term they themselves would choose.

1.7 The Coping Strategies Adopted by SSF Globally

We turn now to the coping strategies that SSFs globally have adopted to deal with the above challenges. Although Jentoft et al. (2011) state that SSF are not as resilient as some people believe, many SSFs around the world have dealt with the above challenges (Eide et al. 2011) by devising innovative strategies (Chambers 1989; Jóhannesson et al. 2003; Clay and Olson 2008). These plans are framed by SSFs' circumstances and the options available to them within the communities they live in (Salas et al. 2011), though not all the strategies are possible or sustainable without external intervention (Henry and Johnson 2015). The most prevalent coping strategies adopted by SSFs globally are improved marketing tactics; diversification; informal and family networks; fishers' organisations; legal and social challenges to the authorities; co-management; migration; asset stripping; and illegality.

Penca (2019) points out that many SSF improve their marketing tactics by promoting the local nature of their products - for example, by arranging gastronomic demonstrations in public and organising local fishers' markets which sell only SSF fish, and setting up fish box schemes. Pascual-Fernández et al. (2019) discuss the increasing number (especially in the USA and Canada) of direct marketing arrangements or short local supply chain schemes adopted by community-supported SSFs, which cut out the middlemen. Witter and Stoll (2017) describe these arrangements as alternative seafood marketing programmes which increase financial security. Some SSF use traceability schemes to differentiate their niche products from the mass products of industrial fisheries in terms of quality, environmental footprint, local employment, and cultural reference, though Pascual-Fernández et al. (2019) point out that schemes such as MSC are too expensive for many SSFs to access.

The second coping strategy is diversification (Prosperi et al. 2019; Salmi 2005). It is generally established that diversification leads to increased resilience in social–ecological systems (Folke et al. 2002; Henry and Johnson 2015; McConney and Charles 2012; Turner et al. 2003; Krogseng 2016). Fishing is a high-risk occupation, prone to unpredictable seasonal and cyclical fluctuations in stock size and location (Allison and Ellis 2001), and diversification reduces the risk of livelihood failure by spreading it across more than one income stream, and by creating opportunities for human agency (i.e., entrepreneurship) (Haque et al. 2015). It also helps to overcome the uneven availability of assets (principally labour) caused by seasonality; to generate financial resources in the absence of credit markets; and to hedge against market failures (Allison and Ellis 2001). Salmi (2015) uses the term 'post-productivism' to depict diversification. Inter-sectoral diversification refers to cases in which fishers become involved in other economic sectors while remaining connected, though less involved, with the fishery (Jentoft and Chuenpagdee 2018). Tourism is a popular complementary activity noted in the literature as a coping strategy among SSFs (Salas et al. 2011) followed by aquaculture, agriculture, and the building trade. Alternative livelihoods are not, however, always attractive strategies because for many fishers, fishing is not only a way to earn an income but a way of life (Kraan 2011; Johnson and Pálsson 2015). Being a fisher is part of one's identity, and fishers may be very reluctant to seek other jobs (Symes et al. 2015; Islam 2011; Jentoft et al. 2011; Pollnac and Poggie 2008). Morgan (2016, p. 1766) says "concepts of self-identity, social role and household status play an influential role in the decision to diversify. On this basis, diversification is anticipated to remain a minority strategy". Intra-sectoral diversification is diversification within the fishing sector itself (Henry and Johnson 2015). Given the high uncertainty of their working environment, some fishers target many species to buffer the effects of a price collapse or reduced catch of a particular species (Berkes 2007; Henry and Johnson 2015). This versatility allows them to gain sufficient income from fisheries throughout the year and helps spread out the pressure on fisheries resources, enabling protected stocks to recover (Chuenpagdee and Juntarashote 2011). Other intra-sectoral strategies include downsizing: for example, replacing trawlers with outboard-engined smaller boats.

A third coping strategy adopted by SSFs globally is informal networking whereby fishers draw on social and family networks (i.e., fellow fishermen, relatives, and patron-client networks) to deal with their problems. Acheson (1988, p. 2) notes that "survival in the industry depends as much on the ability to manipulate social relationships as on technical skills". For example, some communities deal with lack of credit and threat of the risks associated with money lenders by lending each other money in the form of no-interest loans made informally within their own families and extended social networks (Islam 2011). However, this is a limited solution since the amount of capital that can be mobilised through such sources is generally small. Cooperative working by SSFs has been reported by Salas and Pitcher (1999), who described how fishers form teams to go fishing during the windy season, while others share their different skills on an exchange basis. Other examples of cooperation include the relationships built up between coastal communities and inland communities. For example, fishers can seek shelter within inland communities during the hurricane season, while the people from those communities can go fishing seasonally to the fishers' fishing grounds (Salas et al. 2011). SSFs also come together to agree ways of dealing with temporary problems like price crashes, hoping that a reduction in supply would increase prices (Henry and Johnson 2015).

Family networks may be even more important in keeping SSF afloat, since SSF is frequently a family business. The role of women is especially significant because women make both paid and unpaid contributions to the industry, their families and their communities. Some of these contributions are closely related to the activity of fishing. For example, some women are active, independent participants operating on their own or others' vessels in several countries (Kaplan 1988; Chuenpagdee et al. 2006). Other women support their partners by repairing nets (Thiessen et al. 1992), assisting on their partner's boats, filling bait bags, cleaning fish, or cooking for the crew (Porter 1989). In the past in Scotland, wives of fishermen even carried their husbands through the water to their boats (Nadel-Klein 2000). Moreover, women dominate the pre- and post-harvesting SSF activities, mainly as fish processors, in almost every country (Sharma 2004; Skaptadóttir 2000; Chuenpagdee et al. 2006; Zhao et al. 2013; Yodanis 2000; Nadel-Klein 2000; Fröcklin et al. 2013). Other female contributions are more indirectly related to fishing activity. For instance, many women hold roles in SSF finance and marketing (Fröcklin et al. 2013) or are involved in aquaculture or coastal gleaning (Chuenpagdee et al. 2006; Nadel-Klein 2000; Sharma 2004). Also women increasingly find employment outside SSF, which sometimes makes them the primary breadwinner given the inherently unstable and unpredictable nature of inshore fishing (Gustavsson and Riley 2018; Nadel-Klein 2000; Skaptadóttir 2000; Kleiber et al. 2014; Britton 2012b). Indeed, the burden of coping with the inherent uncertainties, dangers and financial pitfalls of fishing may fall disproportionally on women (Britton 2012b; Fröcklin et al. 2013; Sharma 2004). So the survival of SSF and their communities is not only dependent on men: women may be equally or even more important. At a broader level, we can see an increasing role for women as fisheries leaders. For example, in Finland, women have been elected as presidents of fishing cooperatives (Chuenpagdee et al. 2006). In Ecuador, women are active members of local co-operatives and hold high national level positions (Chuenpagdee et al. 2006). In Scotland, Elspeth Macdonald, previously the deputy chief executive of Food Standards Scotland, has recently been appointed chief executive of the powerful Scottish Fishermen's Federation (SFF). In the EU, the Commissioner for Fisheries has twice been a woman (Emma Bonino in 1994–1999 and Maria Damanaki in 2010–2014). Chuenpagdee et al. (2006) claim that where women are part of the decision-making process, they have been integral in shaping well-established systems of customary governance as well as being instrumental in developing new networks to address changing circumstances. Skaptadóttir (2000) notes how women often prioritise community values (such as maintaining kinship relations and networking between households) by contrast to the individualistic values prioritised by men (such as increasing catch quotas).

The fourth coping strategy adopted by the SSF sector is to establish fishers' organisations (Aryeetey 2002; Salas et al. 2011). By organising themselves into a cooperative, fishers are able to have more control over their product, obtain a wider variety of services, and have greater bargaining power than an individual fisher would have (Kohls and Uhl 2001; Jacinto and Pomeroy 2011). Other SSF organisations enforce self-imposed regulations (Deswandi et al. 2012; Evans and Andrew 2011). Some SSF organisations also take steps to reduce conflict by agreeing on fishing-zone delimitation (Jacinto and Pomeroy 2011; Evans and Andrew 2011). Other SSF organisations have drawn up ethical codes for regulating fishers' behaviour.

The fifth coping strategy of the SSF sector is co-management, whereby communities develop cooperative arrangements with local institutions (Marschke and Berkes 2006; González 2011; Chuenpagdee and Juntarashote 2011). Béné (2009b, p. 254) claims that co-management has become the new paradigm of SSF fisheries governance: "Almost every country in the developing world has now explicitly endorsed co-management or some form of CBFM [community-based fisheries management] as one of its main national fisheries policy objectives". D'Armengol et al. (2018, p. 212) say that co-management is "proliferating worldwide" in response to the challenges faced by SSF. In Wilson et al. (2003) there are several examples of successful co-management regimes across the world involving SSF, though in other cases the picture is mixed: co-management has not always delivered its hoped-for benefits. In a powerful critique of co-management, Davis and Ruddle (2012) argue that far from protecting SSF, co-management is a cunningly disguised way of dragging SSF into the prevailing neo-liberal agenda of modern fisheries management, thereby undermining the cultural values and norms of traditional fisheries communities. Béné (2009b, p. 260) claims that most co-management schemes in sub-Saharan Africa failed to improve governance: "Instead, they simply *modified* the status quo by altering the distribution of power and responsibility between the main fisheries stakeholders" [italics in original]. Defeo et al. (2016) found that while co-management worked well for SSF in many Latin American small-scale shellfisheries, it did not work well for all of them: it is not a universal panacea. Kolding et al. (2014, p. 6) assert that "Co-management in SSFs has a mixed record…

In many cases co-management arrangements are top-down created by governments with non-negotiable objectives". Kosamu (2015, p. 366) reports researchers saying that "co-management in Africa appears to be more of an illusion than an empowerment of local fishing communities". However, defenders of fisheries co-management claim that the main reason why co-management fails is because it is misapplied—power is not shared out equally between the parties. Pinkerton (2003) develops the notion of 'complete co-management' to argue that when properly applied, the concept provides a powerful safeguard for SSF communities.

The sixth coping strategy for SSF is to mount legal and/or social challenges to fisheries management authorities. Scholtens et al. (2019) describe legal intervention as a last resort, but when the law itself is the obstacle, SSF may perceive their only course is collective mobilisation to mount a judicial challenge to the government's policies. Scholtens et al. (2019, p. 335) describe social mobilisation as "Let's Fight":

> When governments are implicated in infringements on the rights of small-scale fishers, closing their eyes to misbehaving third parties, or are clearly in no mood to respond to legitimate concerns, collective mobilisation can be a powerful strategy. For example, governments may ignore historical fishing rights, engage in shady joint ventures selling oceanic resources, facilitate or ignore coastal developments undermining small-scale fishers' resource access, or simply displacing them altogether. In such cases, change may not be effectively pursued through dialogue and collaboration, but rather requires collective mobilisation to challenge the status quo, and the building up of pressure in ways that authority holders feel compelled to respond.

As Scholtens et al. (2019, p. 336) note, some academics support SSF in mounting such social challenges, thereby serving as "activist-scholars or scholar-activists".

The seventh coping strategy adopted by SSFs is temporarily migrating from their communities to find fishing work elsewhere (Islam 2011). Such migration is often triggered by the seasonal and spatial variation of SSF's targeted stock (Allison and Ellis 2001), or it may be caused by managerial decisions to increase spatial regulations to combat overfishing (Thorpe et al. 2011). Another form of migration is for SSFs to emigrate to work in prosperous countries and send remittances to their families back home. In several developing countries, international remittance has become an important component of household income, but temporary migration is not without risk of hostility from host communities, and family breakdown.

The eighth coping strategy is asset stripping, which illustrates the fact that strategies adopted by SSFs are not always sustainable. Coping strategies that are pursued because of a lack of alternatives—i.e., desperation strategies (Marschke and Berkes 2006)—often involve running down productive assets which, while providing short-term security, in the long-run leave people poorer and more vulnerable than they were before (Wood 2003; Start and Johnson 2004). The main form of asset stripping is over-exploitation of natural resources: fishers responding to a diminishing stock or price crashes (Henry and Johnson 2015) by putting more pressure on the fish stocks, thus exacerbating their own vulnerability by depleting those stocks (Islam 2011; Andrade and Midré 2011) and flooding the market. This strategy includes targeting species during their breeding season (Islam 2011), and increasing fishing effort (Henry and Johnson 2015). Another form of asset stripping by the SSF is deferring maintenance on fishing vessels, which means SSF safety could be compromised while at sea.

The ninth coping strategy employed by SSFs is illegality. Some fishers use illegal explosive methods (Nguyen and Flaaten 2011; Islam 2011; Salas et al. 2011; Sathyapalan and George 2015). Others use illegal monofilament nets (Islam 2011). Some fishers use chlorine to force the fish to leave their refuges (Salas et al. 2011). Fishers also remove undersized specimens: catching juvenile prawn and shrimp species is the main occupation of women and young girls in some countries, despite a ban on collecting wild fry (Islam 2011). Another strategy is targeting banned species (Salas et al. 2011). Fishers also fish within no-take MPAs, taking advantage of the lax enforcement of many MPAs (Faraco et al. 2016; Cernea and Schmidt-Soltau 2006). Jentoft et al. (2011, p. 455) observe that "illegal fishing practices have complex motivations and are...often as much about social justice as about criminal justice".

1.8 Conclusion

In this introductory chapter, we have provided a global context for small-scale fisheries (SSF), focusing particularly on the existential threats to it posed by the global forces of neo-liberalism and globalisation. We have also discussed different interpretations of the circumstances of SSF across the world and the controversies they raise. In addition, we have examined the particular challenges faced by SSF worldwide, and their responses to those challenges. One point to make is that although SSFs across the world share many common challenges and many coping strategies, there is always something unique about the problems that any particular SSF faces in a given locality and its responses. As Chuenpagdee and Jentoft (2015) point out, every SSF is unique: SSFs do not exist in a vacuum but must be understood in their own contexts (Jentoft and Chuenpagdee 2018). Another point to make is that most of the chapter has been focused on SSF in developing countries because just as more research (until recently) has been conducted on industrial fishing than on SSF, so more research has been conducted on SSF in developing countries than on SSF in developed countries (Salas et al. 2007; Krogseng 2016). This may be because there is much more SSF in developing countries than in developed countries (Salas et al. 2007), or because SSF are diminishing in developed countries but growing in developing countries (Krogseng 2016; Lloret et al. 2018; Maynou et al. 2011; Kolding and van Zwieten 2011). However, this does not mean that SSF in developing countries are irrelevant to SSF in developed countries. On the contrary, as we shall; see, many of the problems faced by SSF in developing countries are also experienced by SSF in developed countries. As Krogseng (2016, p. 10) puts it: "small-scale fishers throughout the world are subject to the same set of problems". Nor does it mean that SSF in developed countries are less important than SSF in developing countries. Indeed, the smaller the SSF sector becomes in developed countries, the more important it is to examine their circumstances in order to halt their decline, since we may be in danger of losing them altogether along with their communities if we fail to recognize their immense value. Krogseng's (2016, p. 41, 46) findings on SSF in Sweden provide a chilling message:

> The main findings of this study indicate that the small-scale commercial fishing industry in Blekinge, Sweden is at risk of extinction…Should small-scale fisheries be allowed to slip into extinction, society at large will lose vast quantities of knowledge, a unique part of their national identity, and a way of life that cherishes characteristics such as freedom, patience, determination, and bravery.

This is the ultimate justification for the current book—to explain why we should support the survival and prosperity of a uniquely valuable part of England's national heritage – the English SSF.

Chapter 2
Interpretive Framework and Methodology

> *"There is nothing as practical as a good theory" (Kurt Lewin)*
> *"Resilience is a core feature of a fisher identity" (Natalie Ross)*

2.1 Introduction

The first part of this chapter explains the main interpretive framework for this book, which is resilience theory. The second part of the chapter explains the main method of obtaining primary data, which is a qualitative approach involving interviews of key informants (KIs) and focus group discussions (FGDs). Resilience is a concept that has long been used in natural science and is becoming increasingly employed in social science. Bene et al. (2012, p. 8) claim that "Since the 1960s, the concept of resilience has been gaining critical mass in academia. It has now become a central paradigm in disciplines such as ecology, possibly replacing sustainability as the ultimate objective of development. In particular, in domains where issues of shocks, vulnerability and risks are critical…the growing influence of the concept of resilience is particularly prominent" (see also Béné et al. 2014; Davoudi 2012). Resilience theory has been used by several researchers to illuminate the situation of SSFs. Symes (2014) explains how resilience theory originated in 1973 in Holling's understanding of an ecosystem's capacity to adapt to turbulence and establish a new equilibrium, and how other writers such as Berkes, Folke and Adjer extended this theory to social systems, and applied it to coastal fisheries communities with a critical analysis of the attempts by managers to impose rigid controls in a vain attempt to preserve the status quo of fish stocks. These critics argued that mechanistic measures such as quota and effort restrictions enforced to achieve abstract targets such as maximum sustainable yield (MSY) ignored the natural variability of fish ecosystems and imposed a straitjacket on the fishing industry, preventing it from adapting to the changing patterns of fish populations. Likewise, Andrew et al. (2007, p. 235) say that "Rather than focusing on yield or profit maximization, we can better define sustainability in terms of the resilience of a social-ecological system".

© The Editor(s) (if applicable) and The Author(s), under exclusive license to Springer Nature Switzerland AG 2021

R. Korda et al., *Resilience in the English Small-Scale Fishery*,
https://doi.org/10.1007/978-3-030-54245-0_2

However, resilience is a complex concept, with many different meanings (Brand and Jax 2007). Berkes and Seixas (2005) say ecosystem resilience is not about resisting change, but about rolling with it. Davoudi (2012, p. 300, 302) distinguishes between three different meanings of ecosystem resilience: (1) 'engineering resilience' which is closely associated with the etymological origin of the term, 'resilire', which is Latin for 'to spring back': "the ability of a system to return to an equilibrium or steady-state after a disturbance"; (2) 'ecological resilience' which means the ability of a system to respond to a disturbance by bouncing forward to a new equilibrium; and (3) 'evolutionary resilience', which abandons the idea of equilibrium altogether, and means navigating constant change in a process of endless adaptation.

Applying some of these ideas about ecosystem resilience to the subject of human resilience, the present study identifies three types of human resilience, drawing heavily on two books: *Resilience: The Governance of Complexity* (2014) by David Chandler; and *The Neo-liberal Subject: Resilience Adaptation and Vulnerability* (2016) by David Chandler and Julian Reid. In response to the insecurity of neo-liberal economic life, these two books draw a contrast between two modes of human resilience: modernist and postmodernist. The modernist mode, which we call *'adaptive resilience'*, encapsulates the notion that resilience lies in adapting to the circumstances that face us. The postmodern mode, which we call *'transformative resilience'*, encapsulates the notion that resilience lies in transforming the circumstances that face us. Salmi (2015, p. 259, 260) makes a similar distinction:

> Resilience refers not only to adaptation but also to the transformation of a socio-ecological system towards resolving conflicting interests and values…Adaptability reflects the capacity of actors in the system to influence resilience, and transformability is the capacity to create a fundamentally new system when ecological, economic or social structures make the existing system untenable.

In what follows, we will explain in more detail what these two modes of resilience mean, and why they are so important in guiding us through our analysis of small-scale fishers' attitudes towards their circumstances. But before doing so, we note that there is a third response to insecurity that Chandler and Reid mention, which we call 'passive resilience', because it encapsulates the notion that resilience lies in fatalistically accepting the circumstances that face us. We will explain the three modes of resilience in order of activism, beginning with the least active and ending with the most active: i.e. (1) passive resilience; (2) adaptive resilience; and (3) transformative resilience. Jentoft and Eide (2011, p. 3) construct a similar typology when they state "we must learn how people within this sector are *coping*, adapting, transforming their situation" [italics in original]. Likewise, Armitage et al. (2017a, p. 8) refer to "three dimensions of resilience…absorptive capacity, adaptive capacity and transformative capacity". Similarly, Krogseng (2016, p. 13) says "For the purpose of the discussion of small-scale fisheries in this paper, resilience will be understood…as the ability of an individual, community, or system to persist, adapt to change, and transform after and throughout a disturbance or difficult experience". Béné et al. (2014, p. 601) say "these three dimensions (absorptive capacity, adaptive capacity and transformative capacity) are seen as the three critical features of resilience"(see also Bene et al. 2012). Coulthard (2012, p. 4) constructs a four-part typology of fisher agency comprising: 'getting out' [non-resilience]; 'getting by' [passive resilience]; 'getting back' [e.g. illegality—a form of adaptive resilience]; and 'getting organized' [transformative resilience].

2.2 Human Resilience Theory

As Marshall and Marshall (2007, p. 1) explain, social resilience is an important concept for fisheries management in that it helps managers calculate the socio-economic impact of their policies: "Knowledge of the resilience of resource users to changes in resource-use policies can assist in the design and implementation of policies that minimize the impacts on people while maximizing the sustainability of ecosystem goods and services". The passive mode of resilience characterises people who resignedly accept their disadvantaged circumstances as a given and carry on as normal, in spite of the warning signs of an unsustainable future (Mowbray 2017). This response to increasingly straitened circumstances may appear to be a mode of non-resilience or vulnerability to inevitable failure, and indeed this is how Chandler and Reid (2016) characterise it. However, passive fatalism can be a means of surviving, if at a reduced standard of living, in that it is a coping strategy of hanging on. This is illustrated by certain fishers who have lost most of their quota but are determined to continue fishing because it is a way of life more than simply a job or a means of remuneration. Passive fatalism can therefore be interpreted as a mode of resilience, albeit an attitude of resignation in the face of challenging present and expected future events which are thought to be inevitable. Johnson et al. (2014) characterizes this type of resilience as 'survival'. Coulthard (2012, p. 5) connects passive resilience to fatalism: "resigned acceptance of misfortune". The philosophical basis of fatalism is the notion that we are powerless to do anything other than what we are actually doing. The notion that we have no power to influence our future is a belief very similar to pre-determinism or defeatism, which prescribes that acceptance is more appropriate than futile resistance to inevitability. As Taylor (1962, p. 56) puts it, a fatalist thinks they:

> cannot do anything about the future. He thinks it is not up to him what is going to happen next year, tomorrow, or the very next moment. He thinks that even his own behavior is not in the least within his power, any more than the motions of the heavenly bodies, the events of remote history, or the political developments in China. It would, accordingly, be pointless for him to deliberate about what he is going to do, for a man deliberates only about such things as he believes are within his power to do and to forego.

However, passive resilient fishers do make a conscious choice in one sense—not to abandon fishing for an alternative livelihood. As Coulthard (2012, p. 5) puts it: "if that alternative livelihood is not a valued way of life for the fisher, and if exiting the fishery would mean 'letting down' his crew members and other dependents in the fish chain…or involve a loss of prestige or social position, he may be unlikely to choose to adapt in this way". TBTI symposium panellists say "If small-scale fishers decide to stay, rather than leave their occupation, it is...because this is their preferred livelihood, and a viable way of living" (TBTI 2017, p. 10).

Turning to the adaptive mode of resilience, Jentoft (2019, p. 313) notes that the notion of adaptation has become hugely popular: "The meteoric rise" in use of the term adaptation...can be seen in concepts like 'adaptive system', 'adaptive networks', 'adaptive capacity', 'adaptive learning', 'adaptive governance', 'adaptive management', 'adaptive co-management", 'adaptive cycle', and so forth. 'Adaptive' is about to become as popular an adjective as 'sustainable'". This mode of resilience is one in which individuals do not passively accept the adverse circumstances in which they find themselves, but take active steps to adapt to those circumstances. Chandler (2014, p. 5) refers to this mode of resilience in terms of "responding ('bouncing back') from disaster or crisis"; "a process through which crises make us stronger, more flexible, and more open to new opportunities"; "about how we can act...to minimise the effects of crises". As Chandler notes (2014, p. 6, 46), this mode of resilience, which he calls the classical or modernist mode, focuses on "the subject's internal capacity to withstand pressures or stresses which were understood to be externally generated...The etymological roots of classical understandings of resilience are framed in terms of the inner resources and capacities of the autonomous individual...to reflexively engage with a complex world". Chandler and Reid (2016, p. 4, 5) refer to the adaptive resilient as:

> a subject which accepts the necessity of the injunction to change itself and adapt in order to cope responsively with the threats and dangers now presupposed as endemic...having to accept that it is not possible to resist or secure him or herself from difficulties encountered (both individually and collectively) but instead learn how to adapt to their enabling conditions via the embrace of insecurity and unknowability.

Adaptive resilients have come to the conclusion that it is not possible to exert control over external forces such as globalisation, and that any attempt to do so is not only futile but hubristic in that they will end up worse off than if they adapt to it. Chandler (Chandler and Reid 2016, pp. 39–40) reported that Giddens argued that "today we have become aware that the aspiration of controlling and shaping our external world was a product of human hubris and misunderstanding. Today's globalised world is a dislocated, uncertain, insecure, 'runaway world'". As Beck (1992) notes, we now live in a risk society, where it is not possible to eliminate risk but only to negotiate our way around it. This does not mean we are helpless in the face of risk, on the contrary, we can be quite creative in adapting to adverse circumstances, especially in changing our own selves, but the focus lies in adapting ourselves to the world as it is: "Resilient subjects are subjects that have accepted the imperative not to secure themselves from the dangers they are faced with, but instead adapt repeatedly to their conditions, in the same manner that living systems...adapt continually to the changing topographies of danger they encounter" (Chandler and Reid 2016, p. 102). Chandler calls this a process of inner transformation: "There is still a discourse of transformation but it is not about the transformation of the external world but the ongoing process of transforming the inner world of individuals" (Chandler and Reid 2016, p. 43).

Reid claims that adaptive resilience is linked inexorably to the discourse of sustainable development in that subjects are required to adapt to an ecological system that is sustainable, and this requirement is particularly directed towards the poor. Ever since the UN Earth Summit in Rio in 1992, improving the resilience of the poor has been a defining objective of ecological health. However, Reid argues that the link to sustainable development turned the adaptive resilience mode into a means of subordinating the economic well-being of the poor to the ecological well-being of the ecosystem. Adaptive resilience, therefore, as he sees it, means forcing the poor to conform to the politically correct green agenda:

> Alleviating threats to the biosphere requires improving the resilience of the poor, especially, because it is precisely the poor that are most ecologically ignorant and thus most prone to using ecosystem services in non-sustainable ways…ensuring the resilience of the biosphere require making the poor into more resilient subjects…[which] requires relieving them of their ecological ignorance…The poor, in order to be the agents of their own change, have to be subjectivised so that they are 'able to make sustainable management decisions that respect natural resources and enable the achievement of a sustainable income stream'… Traditionally, development has been understood to serve the aspirations of the poor to improve themselves economically. Today…poverty is now said by the UNEP to be caused by the poor's dependence on and mal-use of ecosystem services so the aims of poverty reduction have shifted from providing them with economic security to improving their so-called ecological security (Chandler and Reid 2016, pp. 63, 64–65).

According to Reid, this is not a condition of security from the perturbations of externality, but rather a condition of perpetual adjustment to those perturbations:

> The resilient subject of sustainable development is, by definition, not a secure but an adaptive subject: adaptive insofar as it is capable of making those adjustments to itself that enable it to survive the hazards encountered in its exposure to the world. In this sense the resilient subject is a subject that must permanently struggle to accommodate itself to the world. Not a political subject that can conceive of changing the world, its structure and conditions of possibility, with a view to securing itself from the world. But a subject that accepts the disastrousness of the world it lives in as a condition of partaking of that world and which accepts the necessity of the injunction to change itself in correspondence with the threats and dangers now presupposed as endemic (Chandler and Reid 2016, p. 67).

For Chandler, adaptive resilience is thus "a permanent project of self-development, of freeing the subject from their inner limitations" (Chandler and Reid 2016, p. 87). On this interpretation of adaptive resilience, the role of the neoliberal state is not to protect us from risk, but to better equip us to adapt to it. The UK Cabinet Office's *Strategic Framework on Community Resilience* stated that "This programme is part of the Government's 'Big Society' commitment to reduce the barriers which prevent people from being able to help themselves and to become more resilient to shocks" (quoted in Chandler and Reid 2016, p. 43). In the health sector, this strategy of self-help is very clear—instead of demanding more expenditure on the health service for treatment of illness, adaptive resilients make better lifestyle choices, such as reductions in smoking, alcohol, sugar, fat and salt intake, and increased exercise. Indeed, policy-making is not about any particular outcome, but about improving people's capability to make their own adaptive decisions. This means that governmental intervention must be designed not to improve the human condition, but to improve human understanding of that condition in order to better adapt to it (Chandler and Reid 2016).

The transformative mode of resilience rejects the adaptive mode of resilience's focus on the individual's ability to adapt to circumstances, and instead embraces the strategy of changing those circumstances. It dismisses the modernist understanding that resilience is an internal reform of the individual, in favour of the post-modernist understanding that resilience entails an external reform of the system. The modernist mode of resilience is dismissed as an emasculation of human autonomy, by contrast to the postmodernist mode which genuinely frees the human subject:

> the [modernist] process of constructing resilient subjects requires divesting peoples and individuals of any belief in the possibility of determining their own conditions for development and security, and accepting instead the necessity of adaptation to the 'realities' of an endemic condition of global insecurity and to the practice of 'sustainable development' instead…In place of [modernist] resilience…we need to revalorise an idea of the human subject as capable of acting on and transforming the world rather than being cast in a permanent condition of enslavement to it (Chandler and Reid 2016, p. 1, 2)

Chandler claims that adaptive resilients have failed to recognise "that 'freedom' and 'choice' are entirely degraded once the world is reduced to the inner life of the individual" (Chandler and Reid 2016, p. 47). Likewise, Reid argues that an adaptive resilient has "an understanding of life as a permanent process of continual adaptation to threats and dangers that are said to be outside its control…a subject that must permanently struggle to accommodate itself to the world: not a subject that can conceive of changing the world...its structure, and conditions of possibility...[but] a world where humans are stripped of their imaginations, and led to live merely adaptive lives" (Chandler and Reid 2016, p. 53, 184). As Jentoft (2019, p. 305) puts it, "adaptation is not synonymous to development". Jentoft (2019, p. 312) quotes Archbishop Desmond Tutu's remark that "Adaptation is becoming a euphemism for social injustice on a global scale". Coulthard (2012, p. 5) describes this as "the adaptation problem", in that "if people's desires and attitudes are malleable and can adapt to undesirable circumstances, it obscures the way in which well-being can be assessed, and whether adaptation has improved or worsened the quality of life experienced".

By contrast, transformative resilience empowers humans to take an imaginative leap to escape their confinement within the current status quo:

> imaginative action is what enables human beings to forsake the current courses of their worlds in constitution of new ones through, not the transformation of themselves, but the exercise of agency on their worlds...The world thus conceived must conform to the image the subject desires of it and not the other way around…subjects do not merely live in order to fit in with and adapt to existing times, or desire the sustainability of the conditions for their living the lives they do. In contrast they resist those conditions, and where successful, overcome them, transforming them in ways that conform with the transformative work their imagination demands of them; new worlds in succession of old and destroyed worlds (Chandler and Reid 2016, pp. 19, 20)

A transformative resilient is "a subject capable of conceiving the transformation of its world and the power relations it finds itself subject to" (Chandler and Reid 2016, p. 4). Reid says the transformative resilient transcends "its experience of vulnerability, by destroying the very sources of its vulnerability, freeing itself from them, not by living in a state of awareness of their fundamentality to its existence, but by eliminating them, cutting itself off from them" (Chandler and Reid 2016, p. 152). Transformative resilients do not see the solution to economic problems as lying in individual adaptation to the system that is causing the problems, but in collective action by disadvantaged agents to use their knowledge and power to change the system and eliminate the problems (see also Jentoft and Midré 2011). Under transformative resilience, "politics returns to 'the people'" (Chandler 2014, p. 54). So transformative resilience is essentially a bottom-up strategy ("local transformative agency" (Chandler 2014, p. 110) to challenge the top-down policy-making strategy of sustainable development. Chandler says "the problems are in the world, not in our heads...[we need] to remake the world rather than to remake the human" (Chandler and Reid 2016, p. 169). Likewise, Reid characterises the transformative resilient as "A subject that sees the intolerability of the world as it is presently arranged and demands the seemingly impossible: the creation of a new one" (Chandler and Reid 2016, p. 173). Armitage et al. (2017a, p. 9, 12) interpret transformative resilience as "actions taken to fundamentally shift social-ecological conditions, aiming to improve livelihoods and well-being of certain actors...that result in a qualitatively different system 'identity'".

As we shall see, all three modes of resilience (passive; adaptive, and transformative) are to be found in considerable detail in the English small-scale fishery, though our interpretation of the adaptive resilience mode is less dismissive and negative than that of Chandler and Reid. Reid is very disparaging of adaptive resilience, depicting it as a sell-out to neo-liberalism, demanding that the self loses its radical desire to change the external world, and meekly accepts the status quo, content with wheeling and dealing within that restricted orbit of an unequal and unjust system, gobbling up the crumbs tossed to it by the ruling elite. As we shall see, our interpretation of adaptive resilience is much more ennobling than this, and focuses on the innovative and resourceful way in which fishers can use their ingenuity to adapt to their circumstances. In our view, adaptive resilience is not humiliating capitulation to neo-liberal capitalism, but pragmatic adjustment to a constantly changing world, often following some significant change brought about by acts of transformative resilience.

There are three other points of clarification. First, the lines between passive, adaptive and transformative resilience are sometimes blurred and unclear (Armitage et al. 2017a). Second, adaptive resilience is far more frequent than transformative resilience: as Jentoft (2019, p. 316) notes, "The SSF Guidelines mention 'change' 25 times, and 'adaptation' 12, but 'transformation' and 'revolution' zero times". Third, we acknowledge that resilience is not always a good thing. As Béné et al. (2014, pp. 599, 607) put it, "resilience is neither good nor bad; or, more precisely put...it can be good but it can also be bad...resilience is not necessarily positively correlated with wellbeing: some households may have managed to strengthen their resilience but only at the detriment of their own wellbeing or self-esteem". Béné et al. (2014) criticise proponents of resilience theory for failing to acknowledge that some forms of resilience are bad. We are not guilty of this charge, in that although most of the resilience strategies that we discuss in this book are good, some (such as illegal fishing) are bad. We show there are dark sides to both passive resilience and adaptive resilience strategies in the English SSF. Béné et al. (2014) also criticise resilience theory for inadequately dealing with the issues of human agency and power relations, because its focus is too much on ecosystems rather than on social systems. "In much of the debate on resilience and social-ecological systems, the agency of people is often veiled, focusing instead on the ability of the 'system' to recover from shocks… rather than the choices exercised by individuals within the system, who may, or may not, exert control over the processes by which resilience is shaped…Transformation…but also to some extent adaptation can be held back by power structures" (Béné et al. 2012, p. 12, 28). Again, however, we escape this charge because we discuss at length both the role of agency in terms of actions taken by individual fishers, fishers' associations and fisheries communities, and the way in which power is exerted by authorities in response to the adaptive and transformative strategies adopted by English SSF fishers.

However, Béné et al. do alert us to a potential flaw in resilience theory—that it encourages us to think of resilience as the end or goal of SSF strategies rather than a means to the real end or goal which lies in SSF well-being. Béné et al. (2012, p. 47) say "there seems to be a tendency to 'romanticise' the concept of resilience, and a growing number of non-governmental and/or international development organisations seem to have adopted 'building' or 'strengthening resilience' as the new ultimate objective of development". For example, Béné et al. (2014, p. 598) criticise resilience theory for failing to support the poor: "it is not a pro-poor concept, in the sense that it does not exclusively apply to, or benefit, the poor. As such, resilience building cannot replace poverty reduction". They point out that poor people can be resilient by learning to suppress their aspirations and adjust their expectations to suit their wretched conditions: the poor may 'cope' by cutting their food intake or electricity consumption. Béné et al. 2014, p. 615) claim that "what these chronic poor need is not more resilience, but less poverty and less marginalisation… there is no direct and obvious way out of poverty through resilience. Ultimately, development should continue to be about poverty reduction and wellbeing, not about resilience building". A related flaw is that resilience theory ignores the distributional implications of resilience strategies. For example, a community or group may become collectively more resilient, but not every person in it may become more resilient. Béné et al., (2014, p. 616) argue that "if applied uncritically, a resilience-based approach might end up leading us towards abandoning interest in the poor(est) for the sake of strengthening community or even (eco)system-level resilience".

These are important cautions to us not to assume that resilience is an unqualified good, in and of itself. As Béné et al. (2012, p. 45) say, "Resilience is not a panacea… the concept has its limitations". Coulthard (2012, p. 1) states "enhanced resilience does not automatically result in improved well-being of people, and adaptation strategies are riddled with difficult choices, of trade-offs, that people must negotiate". Béné et al. (2014, p. 599) conclude that "the concept of resilience might have been too uncritically—or too rapidly?—accepted as a new, dominant, paradigm by development agencies and that more work is necessary to increase our understanding of how it relates to poverty reduction." Coulthard (2012, p. 1) concludes that "policies that seek to support fisheries resilience need to be built on a better understanding of the wide range of consequences that adaptation has on fisher well-being". We are not so pessimistic about the value of the concept of resilience, since we feel that most instances of resilience are relatively benign. However, we do take seriously Béné et al.'s (2014, p. 615) warning of the "risk of manipulation" by big business:

> In particular, in the context of climate change, adaptation there is a real danger of misuse, or abuse of the term, as it seems to be increasingly co-opted to accommodate rather than challenge forms of development that are implicated in human-caused climate change and other global environmental problems. In particular, we should not be too surprised if, in the name of their corporate 'responsibility', transnational corporations such as Shell or Monsanto soon propose to fund resilience projects in order to 'strengthen poor communities' resilience against climate change or bio-diversity loss.

As Béné et al. (2012, p. 48) note, this is equivalent to saying "Let's continue business as usual, but make sure that communities are more resilient to the shocks created by our model of economic development". This argument mirrors the disparaging interpretation of adaptive resilience given above by Chandler and Reid. Our interpretation of adaptive resilience is more positive than those of Bene et al., Coulthard, and Chandler and Reid, but we do acknowledge that there may be an element of false consciousness in the concept in that fishers may be ideologically lulled into thinking there is no way the system can be transformed, and so their only option is to adapt to the status quo as best they can.

2.3 Methods of Obtaining and Analysing Primary Data

This book centres on a single case study of the English SSF. Single case studies are controversial in the sense that critics claim we cannot generalise from a single case, but defenders of individual case studies argue that they enable researchers to investigate a particular issue in great depth, through the lens of one instance. A single case study does not pretend to be representative of all other cases: indeed, it prides itself on its uniqueness, but it claims to get to the heart of an issue by its very intensity and detail. As Jentoft and Chuenpagdee (2015b, pp. 728–729) put it, "The case study method is not primarily aimed at providing scientific proof as much as in-depth understanding of the nuances and complexities of social systems, institutions and practices…case studies are means to argue a case, which is what social science is about". In this book, the single case study enables us to drill deep down into the experiences of individual fishers and argue a case for supporting the English SSF. Chuenpagdee (2011a, p. 9) writes that "fisheries are lived experiences, and thus need to be examined in their own contexts, namely through in-depth analysis often conducted in social science research", whilst Kurien (2011, p. viii) refers to the refreshing "voice of the fishers" presented in case study research into SSF. As the Foundational Economy (2018, p. 11) puts it: "We all need to listen to direct experience, especially of those whose voices have not been heard".

There have been many studies published on various aspects of the English SSF but none has conducted such intensive research of the situation of the entire sector as we have by tapping into a rich seam of 'lived experience' of both fishers and regulators across the whole of the nation. We justify the inclusion of many quotations from our interview transcripts in Chapters 4 and 5, on grounds that it is important to read the actual words used by fishers and other stakeholders. As Britton (2012a, p. 10) notes, it is a common complaint of SSF that "nobody listens, we're too small…In order to empower SSF it was suggested that it was important to 'tell your story'- what is the story of SSF, the importance of its existence? There is a need to communicate the power of that vision to create voice and visibility and lead to greater education and awareness". So it is fitting that the authentic, complex and unique voice of English SSF is heard loud and clear in the pages of this book. Chuenpagdee (2019, p. 17) says "the richness and meaning of the sector is best portrayed through thick description and narratives". Nevertheless, we expect our findings from the English SSF case study to resonate with the experiences of SSF elsewhere: "The examples from the case studies show that, despite their complexity and distinctiveness, small-scale fishing communities around the world face very similar challenges" (Chuenpagdee and Jentoft 2011, p. 39). We also considered it important not to censor the fishers' language but to quote their words verbatim, including expletives, in order to capture the authenticity of the way they express their feelings.

The main method of obtaining primary data used for this book was qualitative, principally through key informant (KI) interviews and focus group discussions (FGDs). A qualitative approach was chosen because the research questions outlined in the introduction are essentially descriptive, explanatory, and evaluative. The study sets out to understand peoples' lived experiences from the perspective of the people themselves (Hennink et al. 2011). This recognises that participants' perspectives are unique to them as they are socially and culturally constructed, and context is important when exploring participants' experiences. So rather than focusing on so-called objective facts, the study examines subjective experiences (Snape and Spencer 2003). It is a 'verstehen' approach, where respondents reveal their own perspectives on the research topics (Hennink et al. 2011). This approach was deemed appropriate for the English SSF in order to get as close as possible to an authentic picture of this distinctive fleet. It is argued by Onyango (2011) that such an approach is a good starting point for policy development, because the deeper management understand the mentality of the people they manage, the better their policies are likely to be. However, there is a danger that the researcher's own subjectivity might affect her assessment of the participants' perceptions (Denzin and Lincoln 2008). To avoid this danger, questions asked of KIs and in FGDs, and the data analysis process, were carefully designed to minimise the impact of the researcher on the research process and to enable neutrality to be maintained.

In an attempt to provide a comprehensive picture of the English SSF, KI interviews and FGDs were carried out in all ten Inshore Fisheries Conservation Authority (IFCA) districts in England (see Fig. 2.1). The IFCAs are responsible for regional fisheries and marine conservation management within English inshore waters. The ten separate districts groups are aligned with county boundaries and extend out six

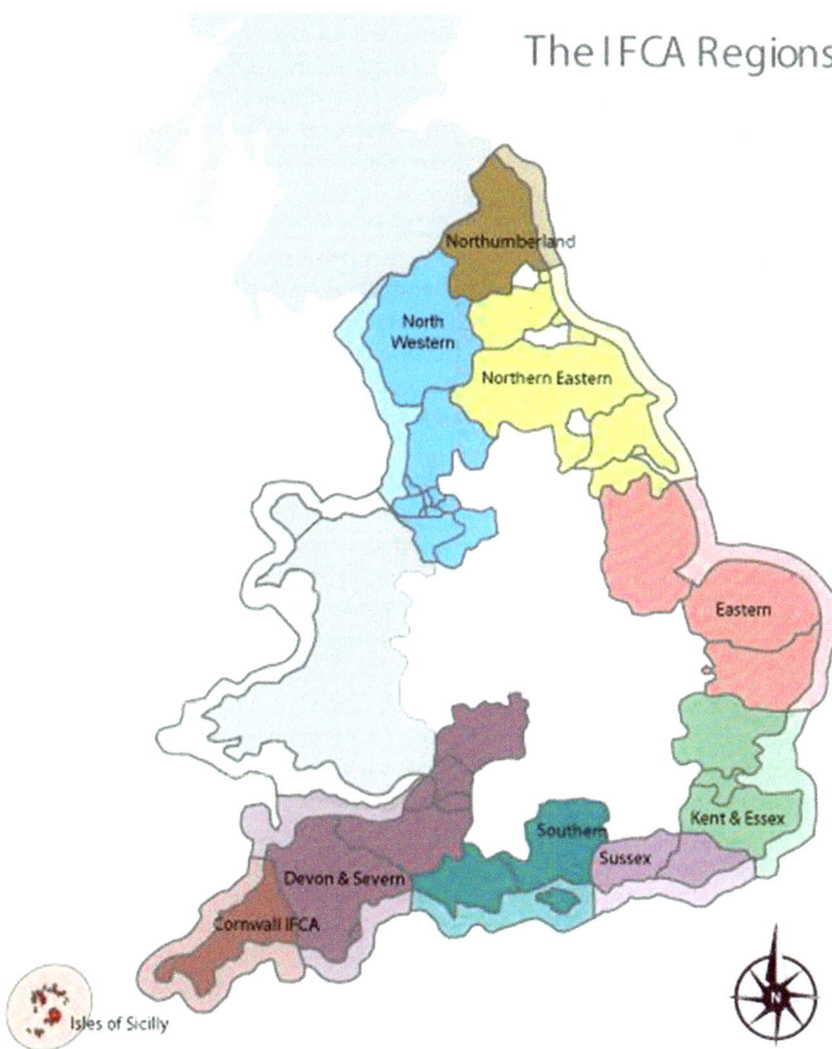

Fig. 2.1 Geographical scope of IFCA districts (Source: The Association of IFCAs, who gave permission to reproduce this map)

nautical miles from their coastline. In order to maintain anonymity, the IFCAs were assigned codes and will be referred to as IFCA 1 to 10 henceforth.

Within each IFCA district, between two and three fishing communities were selected for KI interviews and FGDs. In order to identify these target communities within each IFCA district, the relevant IFCA Chief Fisheries Officer (CFO) completed a score sheet using a Likert scale whereby they could rate different ports or harbours and the fishing communities which operate out of them (which shall hereon be referred to as communities) against variables of resilience along a scale of 1 (disagree strongly) to 4 (agree strongly) as part of their interview. The communities which scored the highest (i.e. deemed by the IFCA CFO to be the most resilient) and the lowest (deemed by the IFCA CFO to be the most vulnerable) within each district were chosen to be included in this study. There was one occasion when two communities got equal scores for resilience, and in this case it was decided that they would both be included in the study. Of course, this strategy presented a degree of risk of bias as it relied on the opinion of each IFCA CFO, which may not provide a true, fair or complete evaluation of a community's resilience. However, the benefits in term of time-efficiency outweighed the risks. A total of 21 fishing communities were selected from the ten IFCA districts.

A 'guardian' was chosen for each community with the help of the IFCA CFOs, the local Fishermen's Mission, the Royal National Lifeboat Institution, and local harbour representatives. The benefits of working through guardians were threefold: (1) in some communities, it was not appropriate to access the community without first seeking the endorsement of a local guardian; (2) guardians were invaluable to glean information about the culture and demographics of community members; and (3) guardians became promoters for the research project within the local community. In most cases, the endorsement of these trusted advocates had a significant influence on whether or not community members participated in the study. Before any KI interviews or FGDs took place, the researcher spent two days in each target community talking to the fishers and informing them about the study. This helped to develop trust resulting in interviews and FGDs that were open and honest accounts of individual experiences.

With regard to recruiting KIs, the guardians helped to obtain most of them, but in addition, opportunistic one-on-one KI interviews were undertaken in each of the communities to counteract the risk of biased samples (Richardson et al. 2005). This approach involved walking around the harbour and the local area, contacting fishers either in their fishing huts, in local cafes, or as they came off their vessels. Also, in order to reach as many inshore fishers as possible, the study was advertised in the popular fishing industry publication, *Fishing News*, local newspapers, and through local harbour authority publications. The Marine Management Organisation (MMO) were approached for an interview, but declined the invitation on grounds of pressure of work. A total of 112 KIs were interviewed, consisting of 88 inshore skippers/owners (13 of whom held the role of fisher association lead (FA lead) and one was also a fish merchant); 10 IFCA CFOs; five FA leads but not active fishers; four local FLAGs (Fisheries Local Action Groups are partnerships between fishers and other local stakeholders which design local development strategies to address their areas' needs and bid for EU funding to carry out the strategies) coordinators; and five local fish market owners. Six KIs were women. Fishers who have left the industry and/or taken early retirement were not interviewed as there was not enough time to track them down within each target community.

With regard to recruitment of FDG discussants, they were restricted to inshore fishers, though on two occasions a FLAG coordinator and a fish market owner turned up and, with the fishers' permissions, were not excluded. Krueger and Casey (2009) criticise creating focus groups from people who know each other, because their well-established dynamics and hierarchies will influence contributions and prevent participants from contributing freely to the discussion (see also Hennink et al. 2011). However, we did not observe this in our FGDs, where pre-existing relationships seemed to aid the process and fishers appeared to be comfortable sharing their views with each other. Focus groups were used as an in-depth group-interviewing technique to gather data on both the attitudes and experiences of fishers, and observe interactions between participants within specific communities (Morgan and Spanish 1984; Teddlie and Tashakkori 2009). This technique also enabled information to be collected from several people in a short period of time. However, despite extensive preparation and publicity to encourage attendance, only 14 focus groups out of a potential 21 took place. In these, the number of participants ranged from two to ten, averaging four participants, and meetings lasted between 52 min and 2.37 h, averaging 1.25 h. All but one of the participants were men. When questioned later, absent fishers explained they failed to attend the remaining seven focus groups for a variety of reasons. Some said the sessions overlapped with long working days and they were too tired to attend; others said they distrusted scientists and managers, while others saw no point in attending a focus group as they could not see how it would be of any benefit to them.

KI interview questions were semi-structured, in order to retain a consistent format whilst giving the researcher space to explore specific topics of interest if the opportunity presented itself. Questions were designed to contribute both thematically to the research investigation and dynamically by promoting positive interaction with the respondent (Kvale 1996). The questions covered five core themes: community culture and networks; challenges facing their fleet; coping strategies used by their fleet; relationships with different stakeholder groups; and managerial efficiency. The interview guide for the IFCA CFOs contained additional questions which examined the CFO's experience, constraints faced by their IFCA, the culture of their IFCA, and their identification of target communities. The interview guide for fishers contained additional questions which examined their fishing practices and fishing backgrounds. The focus group guide was a shortened version of the fishers' interview guide containing 17 open-ended questions, along with a range of pre-designed prompts. Consideration was given to the wording and structure of the questions to avoid leading questions and ensure questions were short, straightforward, unambiguous and clear (Denscombe 2003). It was also decided that a standardized but personable approach to interviews and focus groups be adopted to ensure all participants were exposed to a similar interview experience. Scripts and approaches were tested through a piloting phase undertaken in late summer 2014 with fishers and FA leads across the South East of England attending a conference.

In the main period of fieldwork, each of the ten IFCA CFOs was interviewed face-to-face in their offices between October and November 2014. Interviews took between 53 min and 2.13 h. Averaging 1.10 h. In advance of the interviews, each CFO was provided with an information sheet and consent form. Permission to undertake passive observation of their enforcement and engagement mechanisms was requested, but only two agreed to this, so these observations were used to complement analysis but not as part of the main analysis process. Each transcript was sent back to the relevant CFO to review and confirm their accuracy, and to give them the opportunity to edit out anything they did not feel comfortable sharing. Following this, focus group discussions were undertaken, after which the one-to-one key informant interviews of fishers took place This sequencing was to allow for a range of options to be widely discussed and enable the researcher to understand the target community's norms and values. The focus groups were held in private, comfortable and quiet locations, free from distractions and easy for participants to reach, ranging from rooms in the back of pubs to fishmonger's halls, usually in the evening. Given the often sensitive nature of the subject matter, privacy was crucial as it allowed participants to be as open as they wished. Participants were seated in as circular a position as the location allowed to foster interactive group discussions (Hennink et al. 2011) and level out power dynamics. A dictaphone was used to capture the group discussion as the researcher mainly ran the sessions alone, adopting the role of moderator, and only writing down sparse notes to help facilitate the transcription later—e.g., to ensure that the correct participant was assigned the correct contribution to the dialogue. The main points at the end were summarised to the fishers to ensure that their points had been accurately captured.

Interviews in fishing communities were held at times and locations convenient for the participants (Seidman 1991), usually on the fishers' vessels, but on several occasions upon request, in the privacy of the fisher's home. They were usually undertaken face-to-face, but a telephone interview was offered as an alternative option if necessary. Researchers have generally found telephone interviews to be an acceptable and valuable method of data collection (Sturges and Hanrahan 2004). Initially we considered utilising saturation criteria (Francis et al. 2010) as a guide to how many interviews to conduct in each community (i.e., the point at which the information collected begins to repeat itself). But this was rejected and instead the aim was to interview as many under-10 m fishers and members of the supporting community (such as FLAG coordinators, FA leads and market owners) as possible. Between two and ten individual interviews, averaging five, were held at each location. Interviews lasted between 15 min and 2.29 h, averaging 38.3 min. At the end of the interview with these key informants, the main points were summarised back to them so they could confirm the accuracy of what was recorded in the immediate wake of the interview. This was felt to be preferable to sending the transcripts back in full for them to comment on. With the fishers' permission, interviews were recorded on a digital voice-recording device to allow verbatim transcripts to be produced, thus minimising validity concerns. Only once was permission to record a fishers' interview declined, and in this case responses were recorded by hand, and notes were written up as soon as possible to reduce the risk of mistakes. In all other cases, the dictaphone recordings were transcribed verbatim during the three months following the data collection period. Field notes and a record of observations were maintained to capture interpersonal interactions and non-verbal cues (Gubrium and Holstein 2002).

No financial incentives were provided to any participants as it was felt that this could bias the dataset. However, interviews and FGDs were held at times and locations which were convenient to participants and which did not financially disadvantage them. This generally meant in the evening for the FGDs and a variety of times for the interviews. Information sheets and consent forms were provided to all FGD participants and KI interviewees at the beginning of meetings. These were orally worked through to ensure participants understood the research objectives, how the data will be used, and who will have access to the data. If participants were content, they signed the consent form.

The primary data collected were subjected to qualitative analysis by which the research questions were explored and themes identified, drawing on local meaning and context. Verbatim transcripts of all individual interviews and focus groups were prepared to retain the colloquial style of language and phases used by the participants, in order to retain the flavour and nuance of the participant's original expression (Hennink et al. 2011). Transcripts were anonymised by being given numbers and any personal identifiers removed from the text in order to maintain confidentiality. For example, a quotation from a key informant would be attributed to, say, KI-39 or KI-108, and if the informant were a fishing association lead or a fish merchant or a CFO of a IFCA, or a FLAG coordinator or a fish market owner, that would be indicated. Transcripts were uploaded into the Nvivo 10 programme system (Robson 2011) and a process of data familiarisation undertaken, whereby each interview transcript was explored in detail. Thematic codes arose directly from the transcripts thereby allowing the data to speak for themselves (Glaser and Strauss 1967). In this way, the issues of importance to the participant were captured, which were sometimes different from those anticipated by the researcher. It was decided that frequency of code occurrence would not be included as the number of occurrences does not necessarily signify the value or importance of a theme.

2.4 Conclusion

This chapter has explained the theoretical framework—human resilience theory - used in this research to organise and interpret the findings of the fieldwork carried out into the English small-scale fishery (SSF). We have chosen resilience theory because it focuses on the fundamental issue facing SSFs—how to survive in a challenging environment. We have not ignored criticism of the concept of resilience that it condones survival strategies which confine fishers to straitened conditions. Indeed, we have explained how elements of both passive and adaptive resilience entail acceptance by fishers of unfavourable circumstances. But we have also shown how other elements of adaptive resilience enable fishers to innovate in positive ways to improve their situations, and we have explained how transformative resilience enables fishers to rise beyond their circumstances and take control of their destinies. As we shall see, unlike SSFs globally, where adaptive resilience is by far the most frequent response, English SSFs reported more examples of transformative resilience than both adaptive and passive resilience put together. The chapter has also outlined the methods used by the researchers to obtain primary data. The next chapter describes the historical background of the English SSF, and the following two chapters present stakeholders' perceptions of the challenges they face and their responses to those challenges.

Chapter 3
The English SSF Fleet

> *'In high income countries, small-scale fisheries are at risk of going extinct' (Kyla Krogseng)*
> *'Commercial fishing by under 10m or 'inshore' boats in England...despite a long tradition...has become economically marginal' (Matt Reed et al.)*
> *'Inshore fisheries have for long been a neglected area of policy concern in Britain' (David Symes)*

3.1 Introduction

There is a specific definition of SSF in England which rests on a length threshold of 10 m and under (Davies et al. 2018). This reflects the practice of the English fishery which has long divided the fleet into two parts—the under-10 m (U-10 m) sector (SSF) generally encapsulated by the term 'inshore' fleet' (Sowman 2006; Schumann and Macinko 2007; Hauck 2008; Gray et al. 2011; Davies et al. 2018); and the over-10 m (0–10 m) sector generally encapsulated by the term 'offshore' fleet or 'industrial' fleet. This is an arbitrary definition, and not one that is often found outside the UK. The EU Common Fisheries Policy (CFP)'s definition of SSF emphasises the need to define the sector's vessels by their environmental impact as well as their length, and so identifies them as under-12 m vessel length using low impact gear (i.e. trawls are excluded) (García-Flórez et al. 2014). The British government has suggested that after the UK's withdrawal from the EU, the distinction between inshore and offshore vessels will be removed because it is anomalous, since some inshore vessels (the so-called 'super-under-10 s' or 'rule beaters') are high-powered, technically sophisticated vessels, capable of catching larger quantities of fish than some offshore vessels. Some commentators have said that the inclusion of these vessels within the under-10 m category is completely inappropriate and undermines the category (Davies et al. 2018; Crilley and Estaban 2013). Critics argue that the super under-10 s damage the livelihoods of the remaining inshore vessels by taking

© The Editor(s) (if applicable) and The Author(s), under exclusive license to Springer Nature Switzerland AG 2021

R. Korda et al., *Resilience in the English Small-Scale Fishery*, https://doi.org/10.1007/978-3-030-54245-0_3

most of the sector's quota (Davies et al. 2018) since they are engineered to have the greatest possible catching capacity within their size bracket and they are able to operate in conditions which would be impossible for more traditional inshore vessels. As Symes (2002, p.14) says, "Such vessels threaten the spirit if not the letter of regulations which seek to protect inshore stocks from excessive levels of exploitation and to reserve inshore waters primarily…for local small boat fishing activity". Defra (Department for Environment, Fisheries and Rural Affairs) (2018, p. 28) has hinted at replacing the inshore category with a 'low-impact' category, which would exclude the super-under-10 s, and "allow us to provide increased fishing opportunities, or lighter regulation, for those involved in low impact fishing activity".

The existence of the super-under-10 m vessels poses a dilemma for how this book deals with the English SSF sector, because they are quite different from the rest of the inshore fleet, and therefore should be excluded from our analysis. However, there are more important arguments in favour of their inclusion. One argument is that their inclusion gives credence to the claim made by super-under-10 m owners that they exemplify a very successful strategy of adaptive resilience. Such fishers who choose to enter the under-10 m fleet through vessel modifications have been referred to as 'rule-beaters', indicating that they have managed to beat the system under which the rest of the inshore vessels languish, and so should be applauded for their entrepreneurial nature in adjusting their business model to fit the regulatory regime. Another argument for their inclusion is that the super-under-10 m fleet are dominated by small-scale family enterprises, highly dependent on a local resource base, fishing mostly within the six nautical mile inshore zone with a localised infrastructure (Symes and Phillipson 1997), labour intensive participation, and a catch per unit of effort much smaller than in the offshore sector (Hauck 2008). These characteristics chime with factors accredited to SSFs globally. Finally, at the time of writing, the UK fisheries administration includes the super-under-10 m vessel within the English definition of SSF. For all these reasons, it was decided that they would be included in this work, though we recognise that English SSFs' claims to be environmental stewards may not cover the super-under-10 m vessels given their high catching capacity.

In this chapter, the English SSF fleet is profiled, its governance system is outlined, and its mode of quota management is described.

3.2 Profile of the SSF in England

Most of the UK's commercial fishing fleet is composed of inshore vessels, comprising 4834 of the UK's 6148 fishing vessels in 2018 (MMO 2018). Of the total UK commercial fleet in 2017, 2602 were English-registered inshore vessels and 554 were English-registered offshore vessels, meaning that the inshore fleet constituted 82% of the English sector (MMO 2018). The most common size class in the inshore fleet is 9.8–10 m (the so-called 'super-under-10s'), compared to the next most numerous size class of 4.8–5 m. (Davies et al. 2018). The super-under-10 s also make up a significant portion of the total power and capacity tonnage) of the under-10 fleet (Davies et al. 2018). While different techniques of fishing are used in the inshore zone, 78% of the UK U-10 m fleet's vessels consist of potters, netters, long-liners and rod-and-line fishers (Fleet Register 2016), techniques which are generally thought of as low impact ecologically (Suuronen et al. 2012; Crilly and Esteban 2013). In England, the offshore fleet has a high percentage of towed gear (>90%) (Fleet Register 2016) compared to 22% towed gear in the inshore fleet. The inshore English fleet is operational around the whole of the English coast but the greatest concentration is found along the English Channel; the next largest concentration is located in Cornwall; and the third is located in the North Sea (MMO 2018). According to Huggins (2018), in 2016, the UK inshore fleet landed only 12% of the total UK catch, despite accounting for 88% of the total number of UK fishing vessels. In the UK, the small-scale fleet has a higher number of employees per ounce of fish landed than has the offshore fleet, yet as Percy (2018) notes, the 79% of the UK that is small-scale has access to only about 6% of the total quota.

3.3 Governance of the SSF in England

Current management of the English fishing fleets is undertaken at an international, national and regional level. Internationally, the European Commission sets out the CFP rules which in turn are transposed into national legislation at a Member State level. Within England, Defra oversees the incorporation of EU policy into English national legislation which is implemented and enforced by the Marine Management Organisation (MMO). The Inshore Fisheries Conservation Authorities (IFCAs) are tasked with providing regional fisheries and marine conservation management. Natural England acts as the governmental conservation advisory body to national and regional bodies, and the Centre for Environment, Fisheries and Aquaculture Science (Cefas) provides scientific advice to all levels of management. In addition, the Environment Agency (EA) manages migratory species so they have management responsibilities for the inshore fleet in relation to the prosecution of salmonid fisheries (in particular, Atlantic salmon and sea trout).

Defra's and MMO's jurisdiction stretches out to the EEZ [exclusive economic zone] (200 nm) or the median line (a point halfway between England's and neighboring countries' boundaries), whereas the IFCAs' jurisdictions extend out to 6 nautical miles from their coastlines. The MMO's remit includes managing quota allocations for the inshore fleet; licensing English vessels, licensing marine developments, administering and overseeing fisheries-related grants, applying technical regulation measures to fishers, enforcing the regulations of the Wildlife and Countryside Act and the EU Habitat and Bird Directives; and establishing a marine zonation plan for English territorial waters.

Each EU Member State is able to impose additional marine and fishery management measures insofar as these apply only to their fishermen, or with regard to local stocks prosecuted by other fishermen, and provided they go beyond the minimum requirements laid down within EU law (Phillipson 2002). Until April 2011, this was undertaken by 12 regional management bodies, the Sea Fisheries Committees (SFCs) which had been in place since the late nineteenth century and covered both Welsh and English waters. Under the Sea Fisheries Regulation Act (1966) each of the SFCs gained a statutory-defined remit reaching out to six nautical miles from the mean spring low water mark. The SFCs were responsible for fisheries regulation, stock enhancement, monitoring and enforcement of their own regulations within their districts. Regulations could be generated through the creating of by-laws or fishery orders (but this was limited to molluscan and crustacean fisheries) which could take the form of a several order (to limit public rights of fishing) or a regulating order (to develop licensing systems). These regulations were decided upon by a voting committee composed of representatives from the different sectors active within the SFC's district. In 2011, under sections 153–158 of the Marine and Coastal Access Act 2009 (MCAA), these 12 SFCs were replaced by ten Inshore Fisheries Conservation Authorities (IFCAs), operating solely in English waters (the Welsh SSF are now managed by the Welsh devolved government) (Pieraccini and Cardwell 2016). The IFCAs still operate within the six nautical mile limit (Phillipson 2002) but their geographical remit was aligned to fall in line with county boundaries and includes estuarine areas which were previously managed by the Environment Agency. The switch also extended their duties beyond fisheries to include a wider management of the marine environment (Defra 2010; Pieraccini and Cardwell 2016). Symes (2002, p. 86) points out that even during the 1980s and 1990s, the SFCs were given responsibilities to take account of environmental matters in the discharge of their existing duties, though no new funding was provided to finance these extra responsibilities, and there was little evidence that SFCs took a pro-active role in environmental issues but remained "fully committed to their primary objective of managing local fisheries for the benefit of the local industry". This all changed with the decision in 2012 by Defra to revise their approach on how fisheries operating within European Protected Sites were managed. As part of this decision, they included fisheries under the definition of a 'plan or project', which meant their activities within European marine sites had to be managed in line with how other marine industries were managed, and were assessed for their impact on the features that were protected under European legislation, and suitable action would be taken for any adverse impact. This monitoring and enforcement work is spearheaded by the IFCAs, so in addition to their previous fisheries management duties, they have a heavy programme of work imposed on them by Defra which is conservation-focused, and comes with tight deadlines. This duty has come to dominate the work of IFCAs activity during the last 8 years.

Moreover, whilst IFCAs can set local by-laws, these by-laws are not effective until confirmed by Defra (Appleby and Jones 2012; Pieraccini and Cardwell 2016). Thus the IFCAs are not fully autonomous in their agendas. The ten IFCAs together have about 140 paid staff who manage a total area of over 28,000 km^2 sea area out to six nautical miles around the coast of England with a combined annual budget of nearly £9 million (AIFCA 2015). Like their predecessors, the SFCs, each IFCA is governed by a management committee, the membership of which is partly determined by local councils and partly by the MMO (Pieraccini and Cardwell 2016). Each IFCA must have representation from every local council within their inshore fisheries and conservation area, and these members are democratically elected councillors assigned to the IFCA by the councils. On average, these councillors take up approximately 50% of the seats. IFCA committees must also have representation from the national agencies concerned with fisheries and conservation: the MMO; Natural England; and the Environment Agency (Pieraccini and Cardwell 2016). Other prospective members self-nominate for membership. Unlike under the SFC mode, the MMO is now responsible for this self-nomination appointment process and their selections are named 'MMO appointees'. Prospective members must submit their applications to the MMO who select and interview applicants in line with the rules laid out under MCAA. These rules require applicants to be aware of the needs and views of relevant fishing communities, and to have expertise in the marine environment. Applicants are not eligible for appointment if they have been convicted of a criminal offence and the conviction is not spent for the purposes of the Rehabilitation of Offenders Act 1974. This includes fisheries offences. Critics have commented that the application form is a highly complex, competency-based application that puts off many applicants, and that nominees are subject to a process of vetting by the MMO and Defra in which there is no opportunity for public scrutiny (Pieraccini and Cardwell 2016). An NFFO (National Federation of Fishermen's Organisations) committee criticised the selection process for members of IFCA management committees for the following reasons:

> A centralised application process with decisions on appointments taken well away from the IFCA regions. An electronic process which discriminates against those without access to a computer or who have limited IT skills. Some application questions are tailored more to those who can provide articulate answers to abstract questions rather than those who have the greatest knowledge and experience of fisheries in their districts. There appears to be little constructive feedback for applicants who have not succeeded, which discourages future applications. The process overall gives the impression that the local knowledge and experience of the fishing community is not welcome…In the absence of industry representation, as well as the absence of critical practical knowledge to underpin the decisions of IFCAs, decisions are at risk of lacking balance and having insufficient quorum (Oliver 2016).

In response to these criticisms, MMO said "No appointee is a representative of any sector, organisation or background. Appointments are made in a personal capacity and on the basis of personal knowledge, skills and experience…[nevertheless]… Local knowledge and experience of the fishing community is not only important, but it is a legal requirement to have on all 10 IFCAs…The MMO is committed to ensuring that membership of the IFCAs reflects the interests of all sectors, involved in fishing and conservation, including commercial fishing interests", and as a matter of fact, "the proportion of members on the IFCA committees with commercial fisheries knowledge ranged from 22% to 66%, with an average of over 40%" (Oliver 2016). There is a sharp disjunction between the views of fishers and the views of the MMO over the role of IFCA members: fishers see members as representing interest groups who argue the case for those groups (the adversarial theory of the politics of competing pressure groups); whereas the MMO sees members as engaged in a neutral search for the public interest (the consensual theory of the politics of an epistemic community).

Pieraccini and Cardwell (2016) regard the IFCA system as a form of co-management, providing a valuable vehicle for cooperation between government and governed in the English SSF. However, criticism of the IFCA system is made by Rodwell et al. (2014) and Phillipson and Symes (2010) who point to its inadequate resources and its inability to fully integrate fisheries and marine environmental management. In chapters four and five, we report and discuss the views expressed by SSF fishers and managers on the work of IFCAs, in particular the challenges facing different IFCAs and how their responses have affected the resilience of the SSFs within their jurisdiction.

Two fishermen's organisations form another element in the SSF governance system. The National Federation of Fishermen's Organisations (NFFO) represents fishing interests across all four administrations in the UK (England, Scotland, Wales and Northern Ireland), and is well established in government forums. It was established during the 1970s to provide a voice for the industry in light of the UK's European Community membership and the emergence of a Common Fisheries Policy (CFP) (Phillipson 2002). The financial resources of the NFFO are collected by member organisations and are based on subscriptions and levies on members according to vessel capacity. Whilst it represents both inshore and offshore fleets, its membership is dominated by the larger vessels: 98% of its membership is from the industrial sector by gross tonnage (Greenpeace 2013). Symes (2002, p. 15) comments that "The apparent invisibility of the inshore sector within national policy deliberations is to a degree matched by the under-representation of the sector within the industry's own national political organisation...(NFFO)". The second national representative organisation is the New Under Ten Fishermen's Association (NUTFA), which exclusively represents the inshore sector. It was formed by inshore fleet fishers following the introduction of the Registration of Buyers and Sellers (RBS) legislation in 2005/6, which for the first time provided government with an indication of the catches taken by this sector. NUTFA is much less influential than NFFO, partly because it is more fragmented and has far fewer members and resources.

There are also local port-based Fishermen's Associations (FAs) which aim to represent the interests and welfare of their members in consultation with government and other organisations involved in the marine environment (Phillipson 2002). FAs are very diverse in terms of membership numbers, interests and organisational structures. For example, while some include members who target a wide range of different species, vessel sizes and fishing methods, others have a more specific membership base consisting of a single species or vessel group. The means of generating funds for FAs are varied, including regular subscriptions per vessel or person, one-off payments, levies on landings, and occasional contract work.

Finally, there are fish Producer Organisations (POs), which were introduced in the early days of the CFP (Anbleyth-Evans and Williams 2018). POs are associations of fishers and vessel owners established under EU legislation with financial aid from the EU in the early 1970s (Cardwell 2012). Their role was initially to adjust supply to market requirements in order to guarantee a fair income to producers and stability of supply (Phillipson 2002). This remit changed in the 1990s, and became quota focused. Most PO members are from offshore vessels (Anbleyth-Evans and Williams 2018), but not all industrial fishers have joined POs, and the gradual evolution of quota management saw UK offshore boats split between two different management regimes: the PO members became known as "the sector"; whereas the vessels still managed directly by the government became the "the non-sector". The UK fleet can, therefore, be divided into three separate groups: (1) vessels belonging to a PO which manages their quota management system; (2) vessels over 10 m in length not belonging to a PO; and (3) the under 10 m vessels. Groups (2) and (3), collectively known as the 'non-sector', are managed by the MMO under the direction of Defra. However, a fourth category was introduced in 2015—the Coastal Producers Organisation—registered with the Financial Conduct Authority and based in the Cooperative Futures office in Gloucester. Any inshore fisher can join the Coastal PO, and its aim is "to give inshore fishermen more control and a stronger voice in fisheries policy…as well as better marketing of their fish and bulk purchasing of supplies" (Oliver 2015b). In 2017, the Coastal PO was officially recognised by the MMO as a PO, thereby entitling it to EU funding for three years.

3.4 Quota Management of the SSF in England

Quota management is the pivotal governing instrument in UK fisheries, and judged by the amount of disgruntlement expressed in the fishing media against the unfair way it is perceived to affect the English SSF fleet, we expected it to be overarching issue in the minds of our respondents. However, as we shall see, the problem of quota distribution turned out to be a symptom of a wider grievance voiced by SSFs—lack of participation in fisheries management decision-making. Once the overall share of total allowable catch (TAC) at an EU level has been agreed by the Council of Ministers, it is split amongst the member states in accordance with the principle of relative stability (Anbleyth-Evans and Williams 2018). Until 2012, Defra was responsible for individual quota allocation of the UK's share of the annual TAC set by the EU to fishers in England, Scotland, Wales and Northern Ireland. However, a concordat was signed in 2012 which devolved power to all four of the administrations to manage quotas and vessel licensing (Defra 2012; Smith 2013), allowing each separate nation's administration autonomy over the detailed arrangements that are most appropriate for the vessels under their management. In England, the duty of distribution is held by Defra which has passed it on to the MMO for implementation (Defra 2012). In England, before 1 January 1999, these quotas were attached only to offshore vessels' licenses based on track records which were calculated on the vessel's catch history during a three-year rolling period documented in a logbook (Gray et al. 2011; Phillipson

2002; Cardwell 2012; Appleby et al. 2018). However, by their own admission, many vessel owners were imaginative when logging their entries and this meant that their track record was exaggerated (Gray et al. 2011). In the case of sector vessels, a PO would receive an annual quota allocation based on an aggregate of their member vessels' track records (Phillipson 2002) and distribute the quota as it deemed fit. In 1993, the Government granted POs the permission to trade quotas, and in 1994, to retain the track records of member vessels when their owners surrendered their licences (Cardwell 2012). This meant that an unofficial market for quota grew throughout this period (Cardwell 2012). In the case of non-sector vessels, the quota was distributed by the Ministry of Agriculture, Fisheries and Food (MAFF, the predecessor of Defra) (Cardwell 2012). The number of vessels in the English offshore non-sector (i.e., non-PO) has dwindled to 442 of the 1229 large-scale vessels flying the UK flag and by 2015 approximately 94% of quota by tonnage was under PO control (Appleby et al. 2018).

In an attempt to simplify the quota management system and prevent fishers over-declaring their catches in order to inflate their track record and acquire more quota (Cardwell 2012), Defra introduced a fixed quota allocation (FQA) system in 1999 (Cardwell 2012). The rolling track record was replaced with a fixed number of 'quota units' based on catches made between 1994 and 1996 (Appleby et al. 2018; Hatcher and Read 2001; Davies et al. 2018). It also stated that no new licences would be issued, so anyone wanting a licence had to obtain it by transfer or purchase from another vessel (Phillipson 2002). In 2002, it was decided that FQAs could be separated from vessel licences though only in limited circumstances such as if a vessel sank or was decommissioned: they could not be removed from an active licence (Cardwell 2012). Despite the government's assurances that FQAs were non-transferable and were not to be treated as possessions, the FQA system made quota trading much easier (Appleby et al. 2018), and an unintended consequence of this policy was the rise of the private fishing rights market as these quota allocations acquired a market price. The result has been a concentration of quota within large fishing corporations (Sumaila 2010; Anbleyth-Evans and Williams 2018).

In the early 1990s, whilst this period of regulatory overhaul in the offshore fleet was taking place, many offshore vessel owners rushed to claim licences for inshore vessels, which were not subject to regulatory control, modifying their offshore vessels or using decommissioning subsidies to scrap their vessels and buy or build new ones below the 10 m threshold (NUTFA 2014; Davies et al. 2018). This virtually unregulated inshore sector was particularly attractive to non-sector offshore vessel owners who wanted to avoid monthly limits and a less favourable quota regime (Cardwell 2012). This resulted in a surge in high-catching capacity 'super-under-10 m' vessels, measuring just under 10 m in length (Hatcher and Read 2001; Davies et al. 2018), engineered to have the greatest possible catching capacity without breaking the 10 m length criterion. However, the rush was curbed by the government's decision in 1993 not to issue any new U-10 m licences (Ota and Just 2008), and by two decommissioning rounds for the over-10 m vessels undertaken in 2001 and 2003 to reduce fishing opportunities and withdraw capacity and effort from the

UK fisheries to help stabilise the target fish stocks (MMO 2016a; Cardwell 2012). Unlike later rounds of decommissioning for the under-10 m fleet, which required fishers to surrender their licences, vessels and quota allowances, these initial rounds for the over-10 m fleet removed the vessel and associated licence from an owner, but allowed the vessel owner to keep their FQA allocations, to avoid the government giving them an expensive decommissioning payment for the destruction of their fishing boats. The owners were allowed to move their FQA units to another vessel owner or to POs over the following three-year process, and as a result, a new class of 'slipper skippers' was created, consisting of retired fishers leasing quota out, which had been granted to them as FQAs in 1999, to the highest bidder. Other owners used their decommissioning money to buy 'rule beaters'. Between 1996 and 2006, the number of (active) offshore vessels in the UK fell by 40% (MFA 2009), but a much-criticised hyper-market had been created for leasing quota at exorbitant prices.

As for the SSF fleet, licensing was not extended to its vessels until 1993 (Cardwell 2012), and even then, there was no statutory requirements either under EU or national legislation to record and declare their catches (MFA 2009; MMO 2016a). In 1993, European Council Regulation 2847/93 confirmed that boats measuring under 10 m did not require logbooks, and until 1999, monitoring of their catches was not recorded in a systematic fashion (Cardwell 2012). Instead, data collected composed of landing declarations voluntarily supplied by a few fishermen, figures which greatly underestimated what was actually landed (MMO 2016a). Their catch was also recorded in a randomised manner known as 'stratified sampling' by Defra officials located in the ports, whereby fisheries officers randomly visited a port and took note of every nth vessel's landings. So the figures used to estimate under-ten catches of TAC species were considered to be nominal and minimal (Hatcher et al. 2002).

However, because of the rapid increase in super-under-10 m vessels, MAFF began to tighten control on the hitherto largely unregulated inshore fleet. For example, the sector was brought into the North Sea nephrops (*Nephrops norvegicus*) quota system at the end of 1999, and during the next few years, catch limits for these vessels were extended to all quota stocks (Hatcher et al. 2002), managed directly by the government which allocated quota monthly via a shared community 'pool' (Appleby et al. 2018). Data collected using the aforementioned techniques were taken to reflect the sector's landing ability, and provided the figures from which the fleet was assigned a percentage of the UK's annual TAC. Before this, the inshore fleet was only subject to technical conservation measures such as area closures and gear restrictions, primarily focusing on shellfish species, issued by SFCs (Cardwell 2012). Critics complained that the new quota management regime imposes monthly catch limits which inaccurately reflects their landings on the basis of computer modelling with little sensitivity to regional or annual variations in seasonal catches (Cardwell 2012). Any quota that remains unused at the end of the month cannot be carried over for future use, and inshore fishers are subject to criminal punishment if they break the strict terms of their quota allocation. By contrast, for most of the offshore sector, quota management by the POs allows flexible annual take-up of quota, and individual vessel-owners' conduct is rarely subject to criminal prosecution for exceeding quota allocation (Cardwell 2012). Inshore vessels are allowed to join a PO, but if they do so, they lose their right to fish from the government-administrated inshore quota pool. Of the 4299 inshore vessels in the UK fleet, only 53 are members of POs (Defra 2018).

The introduction of the Registration of Buyers and Sellers (RBS) legislation in 2005 (MMO 2016a; Davies et al. 2018) replaced the random sampling of data collection with a scheme which required the buyers and sellers of first sale fish to submit sales notes (Anbleyth-Evans and Williams 2018). The MMO (2016a) explained that RBS was introduced to enable the inshore fleet to build a track record (which was related to the vessel not the licence), although it is widely suspected that it has been used to make policing of catches easier (Cardwell 2012). The RBS showed that the inshore fleet was landing much more than the government assumed and therefore its pool was nowhere near large enough to account for properly recorded catches (Anbleyth-Evans and Williams 2018). As inshore vessel owners could not remove a quota allocation from the government under-ten pool, they could not take a track record into a PO, so there was no exit strategy for them (Hatcher et al. 2002). Fisheries Minister Richard Benyon admitted that the quota allocation to under-10 m boats was both unfair and unrepresentative of their historic rights (HC Deb, 22 February 2012, c326WH). In 2011, the 78% of the vessels in the UK fleet in the non-sector inshore fleet were given the right to catch only 1.2% of the UK catch in ICES [International Council for the Exploration of the Sea] area IV (North Sea), and only 7% of the UK catch in ICES area VII (English Channel, Western Approaches, Celtic Sea and Irish Sea), the areas where the commercial fish stocks are most numerous. The remaining 98.8% and 93% of fishing rights respectively in these areas, go to the offshore vessels which constitute only 22% of the UK fleet (Cardwell 2012).

In 2007, reacting to criticisms that the allocated quota was too small, Defra presented the inshore fleet with the short-term option of leasing additional quota from the POs at cost, whilst they sorted out the imbalance in the system for the under-10 m fleet (Defra 2009; Gray et al. 2011). However, quota leasing is an unregulated arrangement driven by market forces, so many inshore skippers found that when they needed it most it was far too expensive for them—e.g., £800–1000 per ton of cod in 2016 and 2017, when the resulting catch might bring just £1000 at market. Leasing also increased high grading: because it encouraged fishers to keep only the highest quality fish to pay their leasing charges and discard the rest. Moreover, the system of quota leasing raised the vexed issue of 'slipper skippers' or 'armchair moguls', who own quota without going to sea, having obtained a commodity that was originally distributed free—a practice described by many U-10 m skippers as 'immoral'.

Two further significant attempts were made by the government to address the disproportionately small quota allocation to the SSF fleet. First, in 2008, Defra proposed to reduce the size of the non-sector under-10 m fleet to match its quota through a £5 million decommission scheme, aiming to target the most efficient 'super-under-10 s' to release quota for more artisanal vessels (Defra 2008). This saw owners surrendering their vessels, licences and quota allocations to national government for redistribution (Gray et al. 2011). However, this decision was heavily criticised for being underfunded (the £5 million budget only decommissioned 50 vessels, resulting in a miniscule quota increase (one estimate was from 50 to 55 kg per month of cod quota in area VIId; discriminatory (arbitrary eligibility criteria); and unfair (U-10 m owners had to surrender quota, whereas over-10 m owners in a previous decommissioning scheme did not). There was also nothing stopping the owners using the money to invest in improved vessels, thus enabling these super-under-10 s to become even more efficient at catching fish. The second Defra initiative split inshore licences into two categories (Defra 2008): full and capped. Skippers were allowed to maintain 'full' licences if they could provide evidence of actively targeting quota species based on recorded landings of quota species exceeding 300 kg in any consecutive 12-month period between July 2006 and January 2008. Skippers whose recorded landings of all quota stocks did not exceed 300 kg in any consecutive 12-month period between July 2006 and January 2008 had their licences 'capped' and could only land up to a total of 300 kg per year of quota stocks (Gray et al. 2011). This measure was designed to prevent the gap left by decommissioned vessels from being filled up by relatively inactive vessels being brought into active service (Defra 2008). It is important to note that previous decisions made by Defra meant that fishers' track record entitlement was and is attached to the vessel not to the licence. This meant if a vessel was sold, the track record went with it, and if a new vessel was built there was no track record that came with it.

This two-tier licensing scheme met with fierce opposition from the fishing industry, who argued that capping licences was unjust because it deprived licence owners of thousands of pounds overnight, and penalised skippers who may well have temporarily diversified their operations to pursue non-target species to take the pressure off quota species—the environmentally responsible behaviour that Defra wanted to encourage. Many fishers took advantage of a loophole to by-pass the two-tier licence by re-registering their vessels under either Welsh or Scottish management. Only after costly and lengthy negotiations was this loophole blocked in 2012 by means of the coastal concordat which shifted quota management to the devolved administrations and gave Defra a clearer understanding of exactly what quota held by a PO belonged to English interests (Cardwell 2012). This prompted the UK Fisheries Minister, Richard Benyon, to announce in 2012 that in 2013, unused quota would be permanently reallocated to the English inshore fleet from English vessels in English POs, though he stated that only consistently underused stocks that were in high demand by the inshore sector would be reallocated. Benyon was clearly sympathetic to the SSF cause, remarking that "I do think there is a disparity between the sectors and I do think there is a job to be done" (Benyon 2013), but his room for manoeuvre was limited by practical and legal considerations. The ad hoc way in which quota trading has developed in the UK means that the government has always had a somewhat ambiguous relationship with the market (Appleby et al. 2018). This ambiguity culminated in a high court battle between the government and the UK Association of Fish Producers Organisations in 2013, which investigated whether FQAs could be considered private property. The high court ruled in favour of Defra by declaring that its reallocation of consistently underused quota allocation was legal (Cardwell 2012), but also ruled in favour of the claimants' assertion that FQAs could be considered as possessions as defined in the European Convention on Human Rights and European Union Charter. The judge held that since the unofficial trade in quota had been officially recognised by the government on numerous occasions through reconciliation; since the government had decided that quota could be transferred separately from a licence; and since FQAs had been allowed by the government to develop monetary value and to be marketed, it had been deemed to be a possession (Cardwell 2012). This ruling meant that any attempted reallocation of FQAs that were not consistently unused—the majority—would fail.

The latest policy shift to affect SSF quota allocations is the landing obligation which was introduced by the EU to eliminate discards. Discarding by both the offshore and inshore fleet caught the attention of the British public in 2010, in part through the Hugh Fearnley-Whittingstall 'Fish Fight campaign. There are a number of reasons for discarding (Kelleher 2005), including insufficient quota allocations for particular species, and capped licences, which have forced fishers to discard when quota for a given species has been exhausted (Gray et al. 2011). Also, economic incentives result in the practice of high grading whereby fish of lesser value are discarded in favour of more valuable ones (Morgan 2013; Gray et al. 2011). Legislation in the form of the EU 'landing obligation' was introduced under Article 15 in the 2013 CFP reform (Council Regulation No 1380/2013) (Veiga et al. 2016). This legislation imposed a total ban on all discard activities to take effect by January 2019 (Defra 2009; Guillen et al. 2018). The no-discard policy raises two main issues. First, the issue of so-called 'choke' species; Guillen et al. 2018; Veiga et al. 2016) is a problem for both inshore and offshore fishers. Choke species have the lowest quota in a mixed-fishery, which restrict the fishing opportunities for other quota species - a problem that can only be rectified if either further quota is leased or the next month's allocation transferred. The EU decreed that TAC adjustments will be made to help avoid the choke species problem, (Guillen et al. 2018), and the European Fisheries Council at its December 2018 meeting made some limited concessions to avoid chokes, including international swaps and additional selectivity and avoidance measures. However, any potential uplift is unlikely to be sufficient to solve the problem faced by those targeting mixed fisheries. The second issue is a logistical challenge faced mainly by the inshore sector, because the landing obligation requires more time spent handling and sorting on board as well as processing at ports and finding uses for the fish they are obliged to land (Veiga et al. 2016). A UK-based practical trial concluded that some ports, particularly the smaller ports, will have problems of congestion and added cost (for staff and transport) to deal with the previously discarded fish (Catchpole et al. 2017; Catchpole and Revill 2008). This challenge is going to bear down more heavily on the under-10 m fleet than on the over-10 m fleet as they are smaller vessels with limited storage capacity (Veiga et al. 2016), and this has implications for the maritime safety of those fishers operating these vessels (Villasante et al. 2011). Moreover, unwanted fish that have to be landed are generally characterised by low economic revenue (Macfadyen et al. 2011), and because the SSF sector is running on small margins, complying with the landing obligation could render their fisheries activity economically unsustainable.

In response to the complexities and restrictive nature of the quota system, some inshore fishers shifted into targeting non-quota stocks, but many of these fishers have had their licences capped as a result. Some fishers with capped licences have found bass (a non-quota species) a highly profitable stock as an alternative to the increasingly competitive crab and lobster fishery, but they now face bass restrictions. A similar picture can be seen with other profitable non-quota species such as mullet, salmon and sea-trout, which are now being restricted in Cornwall and Devon by rules preventing netting in estuary waters. Such developments effectively confine much of the inshore fleet to targeting shellfish, but in 2005 a system of restrictive licencing for activity targeted at shellfish was introduced (MMO 2016a). As part of this system, new reporting requirements were introduced requiring inshore fishers to complete diaries of their daily activities and submit them monthly to the MMO. Also, it is rumoured that Defra plans to introduce a crab management plan which is likely to include a quota system.

This complex story of the fishing quota system in England is seen in the literature as a serious challenge to the sustainability of the English fleet (Hatcher and Gordon 2005; Hatcher 1997; Morgan 2013). A handful of papers focus on the negative socio-economic impact of the quota system on the SSF (Gray et al. 2011; Cardwell 2012; Anbleyth-Evans and Williams 2018). Campling and Havice (2014), Appleby (2013), Mansfield (2004) and Symes (1992) criticise the creeping privatisation of quota. Jentoft et al. (2017b, p. 7) note that "The clash is not only between different ways of fishing and different economic rationalities; it is also a power relationship, where small-scale operators are generally the weaker party". This point highlights the link between disputes over quota and inequality of power between the inshore and offshore sectors. For many fishers, the reason for their unfair quota allocation is because they are politically impotent.

3.5 Conclusion

This chapter has described the characteristics of the English SSF, focusing especially on its mode of governance and its system of quota allocation. These are the central issues at the heart of the book, and they raise important questions about marginalisation and injustice. As we shall see in the next two chapters, they are at the forefront of fishers' perceptions of the challenges they face and their ways of dealing with those challenges. The most persistent complaint amongst English SSF is that they are allowed to play very little part in the decision-making process that rules their working lives. Their second most widespread grievance is that the fish quotas they are allocated are grossly inadequate for them to earn a reasonable living.

In the minds of SSFs, these two issues are interconnected, in that because SSFs are largely excluded from fisheries management, their protests against unfair quota allocations are ignored by the authorities who listen only to the voice of large-scale fishers. As we shall see, political marginalisation is resented not only because it disadvantages SSFs economically, but also because it undermines fishers' sense of dignity and self-worth.

Chapter 4
Perceptions of Vulnerability in the English SSF

There is more to fishing than money (Pollnac and Poggie)
Inshore fishing is more about 'earning a living' than it is about making a profit (Symes)

4.1 Introduction

This chapter turns to the perceptions of vulnerability expressed by English SSF respondents in interviews and focus group discussions conducted during 2015 and 2016. We can divide these perceptions intro two categories: external threats and internal obstacles (Islam 2011). External threats are perceived to come from outside the fleet; internal obstacles are perceived to exist within the fleet itself. These perceptions provide a unique record of experiences that are informative, sometimes deeply moving and occasionally amusing. Many of the external threats facing the English fleet chime with those reported in Chap. 1 to impact SSFs across the globe, such as reduced access to resources because of ocean grabbing (played out through a reduced quota allowance, or being displaced as a consequence of industrial vessel activities, marine industry and the instigation of a conservation focused network of MPAs); environmental degradation (including overfishing); reduced product value as a consequence of operating within an increasingly globalised market; and top-down managerial arrangements in combination with the heightened influence of other stakeholder groups. The internal obstacles include a general lack of social solidarity preventing grassroots unity, which erodes SSF's political strength.

© The Editor(s) (if applicable) and The Author(s), under exclusive license to Springer Nature Switzerland AG 2021

R. Korda et al., *Resilience in the English Small-Scale Fishery*,
https://doi.org/10.1007/978-3-030-54245-0_4

4.2 External Threats

Perceived external threats are divided into three categories: economic; governmental; and environmental. Respondents perceive six economic threats: lack of quota; financial costs; competition from other fishing vessels; encroachment by other marine industries; low prices for U-10m fish; and lack of diversification opportunities. The most contentious economic issue raised by participants is '*quotas or lack of it, that is the main challenge*' (KI-71): '*the under tens don't have enough quota*' (KI-21). A fisher explains that the monthly allocation provided to them by the Marine Management Organisation (MMO) prevents them from generating a viable income: '*if you have a family, you can't afford to fish. You can't get a living out of this job*' (KI-71). KI-47 says that '*£144 a week* [is the]...*maximum you can earn on the quotas allocated...That's every fish you've got on quota available and accessible to you. That's everything, skate, place, cod, sole*'. Fishers say the limited quota virtually prevents them from fishing: '*It's really the quota issues that are the bane of the industry. Because effectively, the under-10m fleet has been banned from fishing*' (KI-43), since '*at the end of the day if you can't catch, land fish...that's you stuffed*' (KI-17). A fishers' association (FA) lead says that '*unless we are given some more quota so we can actually hang on...there ain't going to be no fishing boats. So there will be fish in the ... sea, but no fishermen to catch it. Fishermen are the endangered species, not the fish*' (KI-72). A fisher says the cuts keep coming: '*they had cut my quotas by 60%...so I was left with 40%. I said is there anyone here who would be able to survive if their business was cut by 60% because I can't. And now it's gone further than that*' (KI-77). Indeed, a FA lead says more cuts occur annually: '*they just cut it every year*' (KI-41).

A FA lead explains how the backdoor privatisation of quota came about as an unintended consequence of early decommissioning arrangements which allowed vessel owners to hold on to their vessels' track records: '*the guys who took decommissioning should never have been allowed to keep that quota, so that's how you've got an amount of quota in the hands of the few*' (KI-16). Others blame the way quota was allocated to the offshore fleet (underpinned by records collected by the fishers over many years) compared to the way quota was allocated to the inshore fleet (ad hoc calculations by managers). A fisher uses the vivid example of the southwest England mackerel fishery to illustrate the consequences of this pernicious allocative system:

> mackerel fishing was a lovely fishery. Scottish boats hammered it. They used purse seines and decimated it and finished the fishery. But then they came up with the quota system. However, because the Scottish boats were able to say well look at what we landed, we caught thousands of tonnes down there. So now Cornish hand-liners don't have a mackerel quota! You cannot catch hand line Cornish mackerel 'cause we don't have a quota. The Scottish boats that came down here and ruined the fishery hold the quota. So for Cornish boats to catch mackerel on a hand-line means we have to buy the quota from the Scottish companies that ruined it in the first place (KI-22).

A fisher fears that the new landings obligation (banning discards) will exacerbate the quota problem, because the tiny quota for cod allocated to inshore vessels would make cod a 'choke' species, preventing fishers accessing the quota that they have been allocated for other species: '*it's a case now we are discarding in two trips our whole quota in cod... And that will shut the fishery in 2016, because of the implementation of the demersal discard rule, which is a European rule which we must follow*' (KI-47). The Department for Environment, Food and Rural Affairs (Defra) have approved arrangements to allow inshore fishers to buy, or more accurately, lease, extra quota from quota holders, but leasing sometimes costs more than the fish can be sold for: "*these lease quotas if you can get your hands on them they have become so expensive that sometimes you will buy the quota to catch the fish and when you sell the fish it isn't making any more than what it cost you to get the quota*" (KI-21). Moreover, '*some years they buy the quota and then don't catch the fish so are actually out of pocket*' (KI-56). A fisher (KI-58) explains that the leasing system resets itself annually, so if they have not managed to catch the quota they bought by the end of the year, they have lost the right to target it in the next year. Furthermore, if the MMO suddenly closes a fishery, fishers can lose their entire investment in a quota lease (KI-70). A fisher points out that much of the quota is not even held by fishers but by football clubs and 'slipper skippers' (retired fishers) (KI-69). A FA lead explains that before leasing at a market rate became possible, POs used to give surplus quota to inshore vessels for free, but no longer because leasing is so lucrative (KI-16). A fish market owner regards the whole quota leasing system as immoral – '*I believe that the quota should be a national resource and should be owned...by the government and the people of the country*' (KI-20)—an unethical form of rent-seeking (KI-68). One fisher says that he could never lease something perceived as a national asset (KI-43). A fish market owner warns the leasing system will be the end of the under-10m fleet: '*I think the ownership of the quota by individuals will be the death of the family-owned boats and fishing communities throughout the country*' (KI-20). On the other hand, it could be argued that the low impact vessels in the inshore sector (i.e., excluding the super under-10s) are reliant much more on shellfish than on finfish, and shellfish are not (yet) subject to quota, though there are pot limitation schemes in some districts.

There are major capital barriers to entry into the inshore fishing sector: '*for a newcomer, you couldn't get too much of a start under £50,000*' (KI-78)—especially since banks are unlikely to cover these costs: '*If you go to a bank for that, they won't give it to you. You wouldn't be able to pay it back*' (KI-57). These costs include expensive licences: '*the licence on my boat is thousands of pounds*' (KI-2), as well as professional training courses required before fishers are legally allowed to skipper fishing boats: "*You are talking £500 before you have even earned a day's wage... If you want a bigger ticket you are talking three grand*" (KI-10). Also, costs of running inshore fishing boats are escalating. For example, fuel costs are prohibitive: '*many fishermen don't even make enough to pay their diesel*' (KI-69); harbour charges are escalating (KI-6); and the price of bait has '*gone through the roof*' (KI-17).

Participants explain how the limited areas in which inshore vessels can work are often fished out by nomadic offshore vessels or super-under 10s. Two fishers provide the example of how the Scottish scalloping fleet move around the country shifting from site to site to maximise their profits (KI-6): '*When the scallopers come along, they would just wipe the ground clean and it then wouldn't recover for five years*' (KI-80). A Fisheries Local Action Group (FLAG) coordinator says a similar threat comes from nomadic potters: '*they came in here with a thousand pots and they just worked the ground to death...we went back the following winter and it was terrible*' (KI-66). This is a classic case of Hardin's tragedy of the commons:

> anything that most people have got access to will not be respected. It will be a free-for-all... access to common resource has got to go. What we want is our waters back... and we will look after it...Trouble is that all these trawlers now have got what they call track record and they have been fishing in here for decades so they will claim historic rights...But if we could...limit what and where they could go so...not to have equal access which is what they have got now. As, basically, they just come in and plunder (KI-22).

An influx of part-time fishers is also complained about: '*These new people they are not fishermen and they have full-time jobs*' (KI-2) who often have better equipment than the traditional fishers and can therefore outcompete them: '*The actual fishing communities...The traditional fishing communities...are losing out*' (KI-3).

A fisher says '*there is too much competition for what is out there – oil, gas, aggregates and dredging*' (KI-73). A FA lead refers to the increasing number of offshore wind farm arrays: '*We are being overrun by wind farms down here. They are taking all our fishing ground*' (KI-72). A FLAG coordinator (KI-82) says these developments exclude fishers from grounds they and their predecessors have fished for four generations. A FA lead says these marine developments left fishers '*scratting on bits and pieces that are not worth two rows of sheep shit*' (KI-1). Fishers complain that they are always blamed for harming the marine environment—'*it is always our fault*' (KI-76), but a fishmarket owner (KI-20) says other industries cause much more harm:

> the fishing industry tends to be a whipping boy for whatever goes on at sea. So if there are any problems, particularly with fish stocks, it is always because the fishermen are catching too many fish. That is always the only problem...[ignoring] the fact that if you take Plymouth, they dump thousands of tonnes of spoil from the dockyard through licences granted by...the MMO. And through the Norfolk coast, they dig up the sea bed and put in huge wind farms and thousands of tonnes of concrete and destroy fishing grounds. Then again in the channel, aggregate dredging takes place, again on a massive scale, again licenced by the MMO. This changes the whole nature of the seabed.

Defra is accused of prioritising industrial interests over fishing interests: '*Every other industry, so wind farms, pipelines, whatever, they seem to have a far greater voice than these U-10m fishermen*' (KI-68). For example, marine conservation zones (MCZs) are imposing harsh restrictions on fishing to protect marine ecosystems, yet wind farm developments are allowed to operate within MCZ areas:

> If you are going to set up marine conservation zones, then you shouldn't be allowed to put wind farms in them. They have nothing to do with marine conservation; they are not going to help marine conservation. Putting all those cables along the sea bed is not going to help

> marine conservation. They cause problems instead, because of the static and the radiation that comes off them. So why are they allowed in the sea. If they want to make a marine conservation zone, they want to make it not just for fishermen but for everyone. They won't stop the wind farms, just us (KI-92).

Even more outrageously, large-scale aggregate dredging is permitted close to where fishing activity is restricted in the name of conservation: '*They are trying to stop the little boats fishing, but six miles away they are about to licence a gravel extraction project. And it is something like 50 – 60 million cubic meters to be taken. If it was put in a 6m skip, it would go to America and back.... So they can do all that damage and it don't mean diddly. But then they say to us, sorry fellas, you're damaging the sea bed*' (KI-64). It was perceived that big business is given a carte blanche to do whatever it wants by government: '*We got...bloody dredgers...things that Defra could do something about. But they are big business and they choose not to*' (KI-76).

Many respondents complain of receiving low prices for their landed produce: '*they can't make enough money from the fish that they're able to catch to fund their businesses and keep their boats going*' (KI-109). Several reasons were given for these low prices. An IFCA Chief Fisheries Officer (CFO) says domestic overfishing is one factor in shellfish fisheries due to demersal fishers moving increasingly into the potting sector: '*so many switching from trawling to potting has decimated the prices...as basically there is an over-supply*' (KI-103). Another IFCA CFO says low prices in turn causes more over-fishing: '*they've got to use more gear in order to maintain a living*' (KI-106). Another factor is the lack of local market options: local fishers have no choice but to sell all their produce through a single buyer which does not guarantee high prices: '*It's a dreadful closed shop system*' (KI-56). A FA lead complains that the profit from the fish they landed was siphoned off elsewhere in the production chain: '*the middle man is making the money and he certainly isn't doing the fishing*' (KI-4). The obvious alternative—direct selling—relies on continuity of supply, and inshore fishers cannot guarantee such continuity: '*trying to hold on to a customer base is really hard*' (KI-22). An IFCA CFO says another factor contributing to low prices is cheap imports: for example, cheap lobsters from Canada (KI-107).

A FA lead notes the lack of alternative livelihoods to fishing: '*We have no other industries*' (KI-100). Another says the fishers' highly specialist skill-set makes alternative work a difficult option: '*if you take away the fishing here, from the people involved in the fishing, there is no other outlook for their skills*' (KI-99). A fisher expresses irritation at bureaucrats urging them to diversify: '*they all keep shouting, diversify, diversify but we have done that into angling and other stuff and it doesn't work really. It's not a business, it's just pocket money you are getting. It's very easy for someone stuck in an office shouting diversify, diversify. But you have to do your day job and set up another business in the same time, and it's hard work*' (KI-87).

On allegations of governmental threats, respondents perceive threats both from government in general, and from four tiers of government—EU; Defra; MMO; and IFCAs. On threats from government in general, respondents report deep tension between fishers and government. This is partly because of the large number of agencies involved in the fishing sector—'*The amount of bodies we have to satisfy before we can go out to fish is ridiculous*' (KI-1)—and partly because of the number of regulations governing the inshore fleet produced by the authorities: '*too many rules and regulations. It's just ridiculous*' (KI-87). Two fishers say over-legislation makes it hard for fishers to comply: '*feel like every time we are going to work, we are breaking some law, and we almost certainly are*' (KI-59): "*It can make you a criminal if you get it wrong*" (KI-21). One fisher complains about the increasing bureaucracy they face: '*The paper work I've got for my little boat is enough to fill a study…the red tape has taken over the job*' (KI-94). An IFCA CFO (KI-110) says environmental legislation in particular is growing all the time, and a fisher remarked ironically that '*the only marine mammal that doesn't have a protection order on it is a fisherman…We are an endangered bloody species*' (KI-73). Two fishers criticise the lack of coordination between different fisheries authorities (KI-24). '*None of the different organizations talk to each other*' (KI-86). A fish merchant says '*It's not in joined-up writing. They don't have an overview*' (KI-51). This lack of coordination led to unnecessary duplication of inspections: '*We have had…an inspection from* [the MMO] *and the same day the IFCA will walk in. we will say we have just had the MMO in and they will say we don't care, we are here to do another inspection*' (KI-51). A fisher criticises the lack of accountability of regulatory bodies for their fisheries management decisions: '*who is he accountable to? He isn't accountable to anyone…No one chastises him and says that here he made a serious mistake as a fishery officer*' (KI-50). A FA lead says their unmonitored status allows them to play '*god, judge and jury*' (KI-48). Another FA lead claims that managers fail to explain the scientific basis for their decisions (KI-99), and when they use scientific jargon—'*they always try to baffle you with science*' (KI-74). An IFCA CFO acknowledges the opaque language used by managers: '*So the fishermen… misunderstand what is proposed or what is already enforced*' (KI-108). A fish merchant complains that "*People who make legislation don't think of the unintended consequences*" (KI-51). An IFCA CFO admits they have no resources to monitor the impact of measures on fishers' livelihoods: '*We are supposed to predict the economic impacts and how these affect people's incomes and their livelihoods…but that's usually impossible to do. We just don't have the resources to do the research that will be necessary*' (KI-110).

This leads us to respondents' criticism of governments for relying too much on scientific advice whilst dismissing fishers' opinions: '*They don't listen to fishermen. If we say something is black they will say it is white*' (KI-71). Another fisher says this prevents '*the rich knowledge in the industry to filter into the management making process to its benefit*' (KI-21). A FA lead says academic knowledge is no substitute for practical knowledge: '*she might have studied the science at college and have letters after her name, but she has no practical experience… but we have done it all our lives and they don't take it on board, what we say*' (KI-1). Scientists lack practical knowledge: '*They don't go out on the boats. They don't know how to set the gear*' (KI-9). Scientists are accused of using flawed methodologies. For example, scientists '*go on the same bit of ground…for a two-hour tow and then they count the fish and that's what they go on. You could be four miles this side of it and absolutely full up with fish*' (KI-29). A FA lead explains that this method does not take natural cycles of fish stocks into account: '*the scientists in the MMO go to the same place every year. Never mind that things have changed*' (KI-72). A fisher colourfully characterises the scientists' method: '*They have a million-pound job* [research vessel] *riding around on shitty ground rather than coming out with us on a shitty boat but million-pound ground*' (KI-30). Moreover, an IFCA CFO acknowledges that the scientific data underpinning allocations are old and outdated—'*probably five or six years out of date. So there may have been a problem then, but there isn't now*' (KI-107). For example, '*We're being told that we have fished* [cod] *to destruction, but the cod's all over the place*' (KI-7). A FA lead (KI-1) says the mistrust has discouraged collaboration between fishers and scientists.

With regard to alleged threats from the EU, a recurrent refrain from fishers is that the fisheries are remotely (and badly) managed from Brussels: '*the main problem is that the MMO have to do what Brussels tell them to do and Brussels don't give a shit*' (KI-28). For example, one fisher claims that the EU set unfair fish quotas which Britain has to accept (KI-62). With regard to threats from Defra, a FA lead criticises Defra for its remoteness and for being disconnected with what happens on the ground: '*One of the problems with all these rules being made is that the people who are making them have never been to sea before. They have never seen a fish, they sit behind a desk. They will not come down and speak to us*' (KI-69), despite being the main authority governing the U-10m fleet. A fish market owner claims Defra does not grasp the situation of the SSF: '*90% of the fishermen's problems is that we have people running the industry that do not understand the fishermen's problems*' (KI-79). A lack of time in office of fisheries ministers compounds this problem as they are never in post long enough to understand the fleet: '*We always end up with people who are only in it for a few years and then they are gone again and then you need to train someone else up who disappears within a year or two so there is no continuity*' (KI-21). The turnover also means incumbents are not inclined to rock the boat: '*it's a case of, yer we can spend a couple of years here and we'll try not to fuck it up too much but we're also not going to do anything* [significant] *either and then we can be moved on to somewhere else*' (KI-F-22).

One fisher says that Defra gives less quota to its inshore fleet than does any other EU member state's fisheries ministry (KI-43). Two fishers criticise Defra for lacking the courage to redistribute the quota more fairly (KI-44): '*What they need to do is forcibly take quota off these big boats that have got more than they need...But nobody has the political will to do it...there aren't enough votes hinged on it*' (KI-59). Several fishers view Defra's refusal to allocate more quota to the inshore sector as evidence that it wants to close the fleet down: "*what the government have wanted to do all the way along is to set out a series of objectives and policies to basically bankrupt the fleet*" (KI-21); '*they don't want a fishing fleet here and they certainly don't want a U-10m fleet, they are killing us*' (KI-19); '*It's much easier to keep track of the big boats*' (KI-52); 'if *they can wipe us out it will cost them less money to police the fisheries*' (KI-77); 'S*mall boats are a pain in the butt to them*' (KI-72). A FA lead draws a parallel with the coal mining industry in the 1980s:

"Their outlook for our future is for us to go the same way that the coal industry went...They are trying in every way possible to decimate it" (KI-100).

With regard to alleged threats from the MMO, fishers perceive the MMO as the creature of Defra, with very little autonomy: '*The policy makers, who make the decisions that affect us all, are based in London, in Defra...the poor guys at the MMO that are up there in Newcastle trying to work there with quotas I likened to a clown with one leg, only half an arm to the elbow and one hand with three fingers on it, juggling twenty balls, whilst some bastard prods him with a cattle prod from every angle*' (KI-44). However, the MMO is severely criticised by respondents as ignorant of the inshore fleet's situation (KI-52): '*The MMO are not fishermen, they don't really understand*' (KI-10); '*The MMO's fishery officers don't know what they are looking at. They will frequently ask us what it is that they are looking at...How can you respect an organisation that sends such useless idiots out?*' (KI-60). Fishers question the reasons for absurd quota decisions: '*Like this time of year, we need a bit of sole quota, but we only have 300 kilos. But in December, when we don't catch any sole, we had 12 tonnes*' (KI-60). Also, the MMO is accused by a fish merchant of not understanding its own rules: '*You phone up for licensing rules and no one knows anything about it. You end up knowing more about it than they do and that is so wrong*' (KI-51). For example, another fish merchant claims MMO misinterpreted the rules surrounding a FLAG project: '*Getting what you can and can't claim for, wrong...They told us that they would claw back money if we hadn't spent it all by a certain time, but then we looked up the regulations and found that there was no related clause in any of that. Then they denied they ever said it*' (KI-67). A FLAG coordinator recounts how MMO failed to make decisions on FLAG projects before the deadlines expired (KI-82). Some of these delays have serious consequences: a fisher says '*Two of the businesses that we were meant to be funding have now gone bust, partly as it has just taken so long for the amounts to be paid out*' (KI-83).

MMO is also accused by a fisher of a one-size-fits-all mentality, ignoring local circumstances: '*what gets most people's backs up* [is]*...decisions made at a national level when...they need a local level of thought*' (KI-37). They are also accused of enforcing over-rigid rules. A fish merchant says MMO's rules on licensing prevent the inshore fleet from working flexibly: '*the inshore fleet has to have flexibility... Little boats can't go charging around the oceans...you have to react to what is available in your area...you are fishing seasonally and that requires flexibility...of gear type, flexibility of quota use and when in the year you can have it*' (KI-51). Licensing rigidities also undermine opportunities to diversify outside fishing. For instance, some fishers '*started taking anglers out, but if they keep doing that, they are at risk of their lobster entitlement just disappearing*' (KI-102). A FA lead claims that the MMO changes the rules on quota at short notice without taking consequences into account: '*There were boats at sea at night...getting notifications to stop fishing...at midnight, you can't catch any more skates and any more ray*' (KI-4). Another fisher says this places fishers at risk of committing unintentional infractions: '*when I come home from sea, they might have changed it when I was out at sea, so I may have been accidentally fishing illegally*' (KI-7). A third fisher says MMO does not explain the reasons for these sudden shifts in rules: '*They don't explain why; they just tell you they are doing it. You then just have to follow the rules*' (KI-9).

According to some fishers, MMO fisheries officers appeared to view inshore fishers as criminals: '*Every fisherman is deemed to be a criminal in the process of taking illegal fish*' (KI-47); '*They make me feel like a criminal…we are scared of them, scared of making mistakes. 'Cause they can prosecute us under the Proceeds of Crime Act* [which was] *brought in for drug dealers, ill-gotten gains, so if they wanted to prosecute us, they could take our houses…It's terrifying*' (KI-59). A FA lead says MMO regards inshore fishers as their enemies: '*We are perceived to be the enemy…We are not in a mutual relationship*' (KI-48). MMO '*doesn't give a toss about us and* [XX] *has done his best, mark my words, to mess up the fishermen in this area*' (KI-52). MMO is described by one fisher as an intimidating organisation (KI-59). Another fisher accuses MMO of being aggressive: '*It's not that they are only unhelpful, they are aggressive…We used to go in there and fill in the forms and they would help you. But now they say go away, just go away…we're not here to help you*' (KI-47). A FA lead says MMO looks for infractions at any cost: '*They are just looking for something to be wrong all the time…If they find something wrong then they do you for it…that is the type of people they have got in there now*' (KI-61). Another FA lead says '*You can turn the other cheek so many times and now they are red hot on both sides*' (KI-48). A fish merchant says he no longer trusts MMO: '*The trust has gone…he comes in and shakes your hand and is like "how you're doing?". Then he goes on about how he wants to help the industry and all that crap. But you know he is going to screw you over*' (KI-60). A FLAG coordinator states '*it's got to that stage where they have a complete loathing of the MMO*' (KI-14). Symes (2002, p. 73) states that "Inshore waters around England and Wales…are probably among the most heavily policed anywhere in Europe".

Another fisher accuses MMO of denigrating fishers' practical knowledge:

> I'm the last one of a family of generations of fishermen so all this knowledge is being totally discredited by our government and is being treated as if it is worthless information. But we know it's not and we know that it is absolutely fundamental to what we are doing, as any walk of life you have to be trained and then eventually you can practice in it and you take a huge amount of knowledge with you. For example, someone like me who has been in the industry for 30 years, should be held up as an example of someone whom the MMO should go to learn about the business, but it doesn't happen…Instead, I'm made to look foolish… most of the time (KI-21).

For example, a FLAG coordinator says the MMO ignored the views expressed by fishers in consultations that were held on the issue of brown crab quota:

> there has been discussion about putting a quota on brown crab…I attended a consultation meeting where this was discussed, and at the end of the meeting here was a unanimous view [among fishers] that…We have seen the presentation and we really don't think it will work here and here are our reasons why. And the response from the project manager within the MMO was that, well we will go ahead with it anyway as clearly those attending the meeting just can't and haven't understood. I was like, you know, I really can, and this is massively offensive to everybody…There were many fishermen there with the cumulative experience of hundreds of years at sea…but no, we were all incapable of understanding what they were talking about (KI-82).

On the other hand, two fishers say they find MMO to be supportive: '*Personally I find them quite helpful. Certainly with things like...a recent software update...they were great. I forgot to fill out some forms and they just called me to remind me, and they went out of their way to be helpful it felt*' (KI-40); '*They've always been good down here whenever I've been in there with a problem*' (KI-17).

With regard to threats from IFCAs, they are regarded by many respondents as servants of Defra and MMO. An IFCA CFO and a FA lead affirm that Defra largely dictates their authorities' agendas: '*So although IFCAs were set up to be locally accountable bodies, a lot of our program is dictated by national government now*' (KI-109); '*there is a distinct lack of independence within the IFCA system. Although they are supposed to be tailoring things to make them apply to their district, there is still a feeling within their executives that they have to comply with their workload handed down to them... there is a grave danger that the IFCA will just become a delivery boy for ... Defra policy*' (KI-100). Another IFCA CFO explains that since the replacement of the SFCs by the IFCAs, Defra have shifted their focus towards MPA management: '*Through Defra's change of approach, the IFCA have definitely had to give priority to marine protected areas and that's where all our work, or the majority of our work is focused at the moment*' (KI-105).

A FA lead sees IFCAs as also subservient to MMO: '*They* [the MMO] *are there in the background and you feel like they have the real power*' (KI-41). For example, a fisher points out that membership of IFCA committees is in part determined by the MMO: '*IFCA has no power over the appointment of members: that is in the hands of the MMO... They advertised the places, they sifted the applications, they ran the interviews and they confirmed the appointments. The IFCA had no power*' (KI-40). One fisher explains how this deterred fishers from applying especially since the MMO is an untrusted organisation: '*guess who you have to do you your interview with, before you get to go on the IFCA committee? The MMO actually pick who go on the IFCA committee and I think that is fucking outrageous*' (KI-75). A IFCA CFO acknowledges that the selection process organised by the MMO is extremely arduous and time-consuming, and deters many fishers from applying for posts '*it's a very formal recruitment and selection process and I think for many who might struggle to complete an application form it's off-putting*' (KI-108). A fisher remarks that '*the application can be daunting if you are not used to applying for jobs. It's hard to put down examples of how you have made a difference in the past*' (KI-40). An IFCA CFO notes that fishers are not allowed on the committee if they have breached any form of marine legislation, which leaves a very small pool to choose from: '*You can't move without making some sort of offence if they wanted to push it. It's hard to get squeaky clean people on it*' (KI-103). One fisher says loss of earnings also put fishers off from applying to sit on the IFCA committees: '*all fishermen are self-employed, so every day you give up attending a meeting means you are losing a day's pay...You get plenty of crap days so when the fishing is good you want to be able to get out there and fish. It's a disincentive*' (KI-40). Another fisher says evening meetings would work, but other participants would block such a move: '*these greenies wouldn't be up for that, it would ruin their night in front of their lentil soup*' (KI-43). A FA lead says another disincentive to fishers is the inordinate paperwork which IFCA committee members have to read: '*Two I know, who were on the SFC, have come off now as they said for every meeting, they always get a novel to read. The amount of paper work is horrendous and it is all very well for these other bodies, who have nothing else to do than sit in front of a computer all day, but you can't expect the working fishermen to sit down and put up with all that sort of nonsense*' (KI-42).

Fishers also criticise the way IFCA committees are structured. For example, a fisher explains that '*on a 22-man committee, only two are fishermen*' (KI-102). A FA lead claims this creates an imbalance in the way power is distributed in IFCA committees: '*there is a small group of people who make the decisions and everybody else on there just comes along and nods their head and gives their vote*' (KI-3). Another FA lead says the presence of local authority councillors on IFCA committees contributes to this imbalance, a matter made worse by their perceived lack of enthusiasm and commitment (KI-100). A fish merchant says '*They have to be there for half an hour to claim their money, so they stay that long and leave...They have no interest and they only want their money*' (KI-51). Local councillors are also accused of being ignorant about fisheries: '*Someone mentioned something about demersal fishing and one of the councillors asked "what does demersal mean?". And I thought well, I can understand anyone in the street might not know, but you have been on this committee for three bloody years: surely you could go on line and look at this*' (KI-99).

Another criticism of the IFCA committee model is about the fact that since the IFCAs replaced the SFCs, members have been required to see themselves not as representatives, but as neutral contributors to the public good. Defra guidance lays it down that "individual members of the IFCA are expected to work collegiately and not represent constituencies or interest groups" (Defra 2015, p. 10): '*Sea Fisheries Committees used to be representative bodies with so many fishermen and councillors and they were there to represent their sector. Well it's not like that now. It's just a balanced bunch of people that can thrash out a decision*' (KI-100). A FA lead says '*I used to represent the fishermen. But when you were on the IFCA, you don't represent anybody. You can have somebody from a toffee shop, or a bin man and he's got more say than me...The thing is that people need representation...you need representation from the fishermen*' (KI-2). A fish merchant claims this is an intentional power game: "*I think it's about power. I think that excluding spokesmen for the fishing fleet is about imposing power over them*" (KI-49).

A further complaint is that IFCA committees are not welcoming spaces but make fishers feel inferior: '*one of the lads stood up and asked something and they said, "What the hell do you know? You don't have letters behind your name. You are not a professor or doctor, you are just a fisherman, and what do you know?" You come out of their meetings deflated. You feel like you have wasted their time*' (KI-93). Such negative experiences are not limited to membership of IFCA committees: fishers also speak of feeling stupid when engaging with IFCAs' general meetings: '*we are perceived as local bloody idiots. I mean what do we do? We go to sea, and we swear and we definitely smoke too much, don't give a damn and kill anything, and that is how we are portrayed*' (KI-73). A fisher says '*They are treating us like imbeciles*' (KI-1). Another FA lead feels that attending IFCA meetings is a waste of time because their views are ignored: '*I've only been to two meetings. And nobody took any notice of anything I said*' (KI-3). A fisher says they are ignored: '*I mean you tell stuff to the IFCA and MMO but they never listen to you*' (KI-10). Another fisher speaks of a feeling of futility: '*Cause no matter what you do, you just feel like you are just banging your head against a brick wall*' (KI-23). Other fishers say they are disillusioned and fed up: '*The trouble is that over the years and the decades… fishermen come to all these meetings and nothing changes. And they are disenchanted*' (KI-68). A fish merchant says 'th*ere is no point any fishermen being in there…The only people who are gaining are the conservationists*' (KI-51).

More broadly, some fishers criticise the whole consultation process established by the IFCAs as a façade. A fish merchant says IFCA's CFO '*pre-conceives the outcome and then makes the consultation give him that answer...we were never allowed a free and open discussion on what could be the issues…It was a lash-up from start to finish*' (KI-51). Other respondents hold that IFCA consultations are carried out purely for show: '*It's all already signed and sealed in blood before we hear anything about it... We are going to the meeting tonight, but it will all be settled already*' (KI-93); '*They have these meetings, but I think that these things are already a done deal before they have these meetings. These meetings are just for show. I have been to them before and you just come out of them deflated, like you have wasted your time, because they have made up their minds already*' (KI-92). Fishers allege that IFCAs manipulate information to suit their own agendas (KI-7). A fish merchant says '*they justify what they want to justify; they do not provide you with accurate information…you get punch drunk with* [fruitless] *consultations*' (KI-51). The IFCAs are even accused by a FA lead of downright lying: '*There is a report that is coming out. It's totally misleading, deliberately misleading and it contains a number of lies and it's unrepresentative of what is really there. This has influenced decisions which have had dreadful impacts on a lot of local fishermen*' (KI-3).

As a result, some fishers speak of dissociating themselves from participation processes: '*I never go to the meetings…as it is a waste of time….you just need to accept what they come up with and just grin and bear it. I am not interested in what goes on in those meetings anymore*' (KI-90); '*There is no effective consultation. Fishermen have apathy and can't be bothered*' (KI-68); '*We have been going to meetings and writing letters to politicians for thirty years, and it just goes in one ear and out the other. So I am just at a stage now, where I won't even engage in any kind of political system*' (KI-59). One fisher remarks that '*We used to have a really good working relationship with the fishery officers, but now we see them as our enemy*' (KI-59). A fisher accuses IFCAs of being authoritarian: '*this is what is going to happen…they are telling you, and if you don't like it, tough*' (KI-87). IFCAs are described by a fish market owner as operating from an ivory tower (KI-20). Fishers say fishery officers are invisible: '*You never see anybody*' (KI-95), as are the CFOs: '[the CFO] *never gets off his arse and comes out any more*' (KI-50). A FLAG coordinator complains that IFCA management staff '*just don't want to talk to people*' (KI-82). A fisher claims that IFCA officers '*just want to get on with their paperwork. You can't just go and knock on his door*' (KI-7). Another respondent says '*The chief fisheries officer is supposed to be accessible by the telephone or email…But you ring him up and he is never there. He has a big photograph on the website, saying contact him but it doesn't work*' (KI-84). A fish market owner accuses an IFCA CFO of never returning phone calls: '*I can't phone the office...and ask to speak to him. And they would say every time, he is not available, he is in a meeting…and he will never return your calls*' (KI-79). A fisher says that even when you do get through, nothing seems to happen: '*You can go and say something to XX, but then it doesn't seem to go anywhere*' (KI-54). KI-37 says the officers use unintelligible management-speak. Another fisher complains that information is withheld by his IFCA—for example, it did not warn fishers about the imposition of a MCZ: '*I only found out about the MCZ on Facebook and it wasn't the IFCA that put it on…They didn't write to us about it*' (KI-98). The opposite complaint is made by another respondent—that the IFCA send out too many messages: '*The trouble is you get so much shit and maybe out of every 50 emails, you might have one important one ... But that one might be vital*' (KI-72).

Behind some of these criticisms of IFCAs lies a perception expressed by fishers that the IFCAs do not have an adequate understanding of the complexities of inshore fleet operations: '*They don't know how the fishermen work, or what they have to do to make a living. It's stupid*' (KI-91); '*They have no experience of fishing. They have no idea what goes on out there*' (KI-74). A fisher says '*The fishery officer we work with ... he didn't even know the difference between a hen crown and a cock crown, or how to measure a lobster, and these are the guys who are supposed to be telling us what to do, governing us, controlling us*' (KI-13).

On the other hand, some respondents are very positive about their IFCAs. For example, one respondent reports that his IFCA is very accessible: '*If we wanted to ask them anything there was that option. You could always ring them up and ask them. They are very approachable 'cause we get on with them*' (KI-35). Another respondent commends his IFCA for being pragmatic in not adopting a purely environmentalist approach: '*On the whole, our IFCA, I feel like the chief fisheries officers and his team are trying to hold back a flood of environmentalists who are putting enormous pressure on fishermen. Their mantra is that the fishermen are clambering all over the seabed and taking every living thing out of the sea. But our IFCA know that isn't true and they do want to get on with everyone. They have a lot of pressure on them from the environmentalists with the conservation zones, but they seem to take a pragmatic take on it*' (KI-41). Another respondent praises his IFCA CFO for his inclusive management: '*I am now on my 4th Chief Officer and this one is the best. He is brilliant ... He gets feedback from people on the ground and he deals with it. And he can deal with bureaucracy upstairs. He is good at whatever he does…He is an exemplar. But how you make the others do that, I don't know*' (KI-39).

On environmental threats, respondents perceive them to come from two sources: the green lobby and the implementation of a network of MPAs. Many fishers express hostility towards the green lobby. An IFCA CFO says '*Environmentalists don't trust fishermen and fishermen don't trust conservationists*' (KI-110). A fisher claims environmentalists have cleverly used the media to vilify the industry:

> I watched that Hugh [Fearnley-Whittingstall] program…he filmed these areas where the scallop vessels supposedly go and…he filmed a sandy bottom. Now everybody knows that scallop dredgers don't work on a sandy bottom. Yet here he was. It was totally false. But then for Jo public…they are going "Crikey, the bottom is totally barren. Look at what the fishermen have done". It gets into people's heads...all the people were like "All the fishermen must be stopped", and once people get like that, you can't get them back...If people all tow that line, it will kill our job (KI-98).

An IFCA CFO says '*fishing has a generally bad press. The general public… questions the right of fishermen to go out and catch fish as a public resource*' (KI-110). A fisher complains that they are blamed for every decrease in stock levels: '*Soon as the stock decreases, they blame it on overfishing*' (KI-29). A fisher says that through the media, '*everybody…has been fed the line that the ocean has been over-exploited and that fishing has caught every last fish that is available and that it has to be stopped and that everything has to be protected*' (KI-21). A FA lead explains how such misleading perceptions are fuelled by stories of small numbers of individual fishers breaching regulations: '*There is an element within fishing that don't have a thought about tomorrow and unfortunately they give the rest of the industry a bad name*' (KI-16)

An IFCA CFO notes '*the greens have a lot of clout now*' (KI-103). A FA lead says '*They have a lot of power these greens. They have a lot of money. All middle class and they have nothing to lose. They just want to try and save the world but they don't get it*' (KI-61). A fish market owner says by contrast, fishers are weak: '*The problem is that the fishing industry on its own doesn't have anything like the power of Greenpeace or the Wildlife Trust...Or the funds*' (KI-20). A FA lead views the greens as an existential threat to the fishing industry (KI-61). A fish market owner alleges that '*These are agendas by people who really don't want to see any fishing at all at sea. And they think that people will exist on beautiful clean seas where no fish are caught at all*' (KI-20). A fisher blames the green lobby for the IFCA's shift from the focus of their predecessors, the SFCs: '*They used to be a Sea Fisheries Committee, but that word Conservation* [Inshore Fisheries Conservation Authority] *is the biggest word in the title now...the problem with them is that the C is bigger than the F*' (KI-73). Two other fisher agree: '*There is no doubt that the IFCAs are very much weighted towards the green element… there's not sufficient fishing representation to counter it. But maybe that is what the IFCAs were designed to do*' (KI-42); '*The green environmental lobby is infiltrated into every body. There may be on the IFCA committee only two or three people from environmental-based groups, but their voice will be so powerful because everybody else has been told that the fish stocks need protecting. Even if they have no idea about fish stocks, they will believe that lie*' (KI-21). A FA lead accuses the IFCAs of adopting this conservation agenda wholesale, forgetting their responsibility for food production (KI-1). A fisher complains "*There is no management strategy within IFCA to help food production and the sourcing of food within the six-mile limit which is what the IFCA are responsible for*" (KI-21). A fisher says the government is afraid of litigation: '*If Greenpeace puts something in...the IFCA has to follow as they are frightened of being sued*' (KI-102).

One of the most visible manifestations of environmental threats is the proliferation of MPAs across the country. The areas where inshore vessels operate are often the focal point of nature conservation. Approximately 23% of English inshore waters are now protected under a network of MPAs driven by a raft of international legislation (such as the EU Birds and Habitats Directives resulting in Special Areas of Conservation, Special Protected Areas and Ramsar sites) (Defra 2012), and national legislation (such as Sites of Special Scientific Interest and MCZs) (Jones 2008)). In 2013, Defra revised the way that fisheries are managed within all European Marine Sites in England in order to comply with Article 6 of the Habitats Directive to ensure that they have no adverse impact on any of the sites' features. This has affected much of the inshore fleet's access to their traditional grounds where their gear has been found to interact negatively with the protected features. The European nature conservation frameworks assess impacts purely on ecological grounds and do not take social or economic impacts into account (Kochalski 2017). A fisher complains that

> They are closing areas now for the future. There is an area just off here, good fishing in the summer, it's local for us, we can get there, we can make a living out of it. But just because of this growth on it, some kind of seaweed, they are going to shut it off [to fishing]. Little bit further down the coast, there's an area that they're going to stop people fishing in too. Completely stop it. Why? Because now and again, birds get caught in nets (KI-7).

One fisher remarks that '*It seems that they want more areas out there that we can't fish than areas that we can fish*' (KI-74). Another fisher says '*They just keep taking and taking and taking. They have taken enough. We want them to stop*' (KI-93). A FA lead says proposed closed areas would force him to fish further offshore into dangerous waters: '*if they close that down then we are going to have to go much further out and...a lot of times it's not safe to go out that distance*' (KI-72).

4.3 Internal Obstacles

Internal obstacles come from within the industry itself. Respondents perceive four kinds of internal obstacles: psychological; educational; social; and physical. On psychological obstacles, an IFCA CFO blames SSF for over-fishing, accusing fishers of opportunist and short-termist behaviour:

> What is fishing about? It's about making money. It's not about preserving stocks for the future. There are very few that have that view...When you say "If you keep fishing like this, there is going to be nothing left for your son to do", the response was "I'm fishing like this, because I never want my son to have to do this"...If it is just a cash cow which you milk until it falls over, then you are not going to be in the game of sustainability; you are going to be in the game of getting rich quick. If you look at the trend of fishing vessels over time and technology, it's how much can I catch, with the least amount of money and the biggest vessel, so one net can gather up everything that is there and make me a huge amount of money. You know that is seen as the efficient business, but it is completely out of step with nature and nature can't sustain that kind of thrashing...Call it what it is, it's greed (KI-106).

Another IFCA CFO says fishers wrongly blame management for their own damaging behavior:

> the fishing community…won't want to hear this but there is a victim status – "It isn't fair, it shouldn't be this way, it isn't right, and it wasn't like this in 1960"…There is a perception of, it wasn't broken, it didn't need fixing. Bureaucrats then got involved to try and fix it and they have wrecked it…they won't entertain the fact that…the very fact that bureaucrats had to get involved is that you were fishing in an unsustainable manner and you were driving stocks out of existence. And there the uncomfortable truth is. But that's not what you hear, "no, no there used to be hundreds of lobsters" ... Well yes, and then you fished them all out, didn't you? That's why they are gone (KI-105).

This IFCA CFO adds that fishers falsely believe they have a God-given right to use the sea because they were there first: '*There is a perception with fishermen that the sea is theirs in the first instance…they were the first operators at sea and therefore they should have preference…I'm a fisherman, it's mine. But it's not…It belongs to the population. You just happen to be lucky enough to have chosen a profession where you go out and experience it and enjoy it daily*' (KI-106).

Another psychological obstacle is fishers' reluctance to serve on IFCA committees. A FA lead says IFCAs find it hard to recruit fishers: '*the right people don't come forward. They struggle to get people*' (KI-2). One fisher says this is because fishers are too busy: '*they won't step forward as they are too busy earning a living*' (KI-64). Another fisher says SSFs are uncomfortable in formal meetings: '*A lot of them are not used to formal meetings and conducting themselves in a formal situation*' (KI-F-40). A FA lead says many SSFs feel intimidated: '*Fishermen get intimidated by a room full of suits. And when those suits start turning up with their lawyers and start bandying things around in official language, or legalistic language, they are intimidated. Any situation outside of your usual work is going to be intimidating. If you stuck one of those lawyers on the deck of a fishing boat, he is going to be intimidated too. It's just taking people out of the world they work into another one*' (KI-81). A fish merchant accuses fishers of shying away from engaging: '*I find them so short-termist. If they don't want to voice their opinion, and some of them have got some brilliant ideas...if they don't go to these meetings then how can they expect for things to go their way?*' (KI-51). One IFCA CFO says fishers have a '*head in the sand*' mentality (KI-106). Two fishers say fishers have only themselves to blame: '*The only barriers are internal and because of our own stupidity*' (KI-39); '*The ones that don't say anything usually shout the loudest against it when it is brought in. Well you were told and you were asked. But we are back to fishermen being their own worst enemy*' (KI-73). A FA lead suggests that this reluctance to engage reflects fishers' suspicion of authority: '*Fishermen by their very nature are suspicious of authority because I think all they ever see is bad stuff coming*' (KI-16). Another FA lead agrees: '*We don't like being told what to do by someone who is sitting in an office and has never been fishing in their lives*' (KI-3). A fish merchant explains that '*Most fishermen are a real dichotomy and they can be really noisy but at the same time, really shy*' (KI-51).

On educational obstacles, a fish merchant notes that inshore fishers often have limited educational achievements: '*not many of them are well educated. And that is why many of them ended up in fishing, as they...didn't feel like they had an alternative*' (KI-51). A FA lead says of one fisher: '*He can hardly read and that is the biggest problem with a lot of fishermen who are...50 and over...I used to do all the writing for them*' (KI-2). Lack of computer literacy is another weakness, because many inshore fishers are not comfortable with electronic means of communicating, and this limits their ability to fulfil their legal paperwork requirements: '*Everything is on line and if you aren't computer literate, then you are in trouble*' (KI-2). Some fishers, especially older ones, do not have computers or smart phones (KI-24): '*Strangely enough, an old bastard like me never learned how to use a computer. I have done it now but I...find it very hard*' (KI-76). There is very little paper correspondence nowadays (KI-10). This means communications can be easily missed: '*They have websites and...sometimes put these consultations on their websites but if you're not computer savvy...you may not even see these consultations....If you miss it on the websites then you don't get to have your...say, so things get passed without you even realising it*' (KI-13). Their educational shortcomings mean some inshore fishers lack the ability to apply for grants. A fisher says: '*we keep getting told about these pots of money from Brussels, but we can't see where the angle is to get at them. Or where they are*' (KI-66). A fisher adds that they have neither the time to complete grant applications nor the money to pay others to do it for them (KI-65).

On social obstacles, the inshore sector lacks fraternal solidarity. A fisher says of the inshore fleet: '*it's not very together, the U-10m fleet...Everyone does their own thing*' (KI-17). Another fisher claims that inshore fishers' communities are splintered: '*They are... too fragmented and don't think with one voice*' (KI-68). An IFCA CFO denies they are real communities: '*the problem with these communities is that they're not communities. They are different parts of the communities... The idea of a fishing community is rather fractured in...many of our developed south coast towns...There are individuals in those places and in most instances it's quite* [divided] *They are essentially...a lot of people competing with one another. So they cooperate only where there is advantage*' (KI-112). One respondent ascribes the lack of solidarity to territorial rivalry and competition: "*in fishing, they all think that they are in competition together*" (KI-F-68). Another respondent refers to a lack of trust between fishers: "*I don't think they trust each other at all. They are all fighting for their piece of sea and they're all fighting for their interests within that piece of sea*" (KI-105). One fisher (KI-94) says this disunity is a recent phenomenon, but a IFCA CFO attributes it to their genetic heritage:

> They tend to be very individualistic. They tend not to show willingness...to get together as a group and ...set up fishing cooperatives...I think that it's ingrained...I think it goes back in people's genes...down the decades and the generations, where they've had to be very independent, they've had to go out and do their own thing really to keep their heads above water to get bread and fish in their families' mouths...and therefore it's a little bit against the grain for them to think in terms of "Oh yer, well if I was to link up with him next door and him in the next port, we could maybe set up something that would give us greater strength moving forward". That's not really in their make-up to do that (KI-111).

Another fisher says he is not a joiner: '*I've never wanted to, never, not once. I always just keep meself to meself. I have never even thought about joining any groups*' (KI-96).

One manifestation of this weak social solidarity is a lack of inshore fishing associations (FAs). A fisher says an FA would never succeed in their area because of disunity (KI-78): '*It would be a help. But fishermen wouldn't agree here*' (KI-77). Where FAs do exist, they are often moribund due to lack of cohesion: '*As soon as somebody falls out with somebody else, it's like a school playground here. So we are not totally united and it's not a good thing, because…lack of cohesion will kill them in the end*' (KI-39). A fisher says that support for his FA is waning and it might be disbanded (KI-90). Another fisher says the lack of FAs is due to dwindling fleets (KI-70). Even where there is a healthy FA, it may well be at loggerheads with other FAs, as an IFCA CFO reports: '*At the moment there are 20 boats but they have three fishermen's associations. And they all hate each other and most of them are related*' (KI-108).

There is also tension between adjacent communities, both in the past—'*when* [community X] *fishermen used to go to* [community Y], *they used to stone them. You laugh but it's true. There was a lot of rivalry. We all wore different coloured jumpers and you didn't wear that colour in that town*'—and in the present—'*we had a gentleman's agreement and that has just gone out the window with these … guys*' (KI-65). According to one fisher, such local divisions render U-10m fishers weak: '*That's their weakness. As they don't stick together and they don't have a single voice, so they don't have a voice. And that's…why their quota has been allowed to be decimated. And that is why with the wind farms…they put them exactly where they want them*' (KI-68). A fish merchant refers to the failure of SSFs nationally to unify: '*we just can't seem to get people to stand together*' (KI-60). A fish market owner agrees: '*Different sections of the fleet have different agendas so the industry has never been united. It's made up of different individuals and different types of fishing*' (KI-20). A fisher explains that SSFs are geographically isolated which makes it difficult to create national unity (KI-10). An example of national friction is the attack made by inshore fishers on an inshore fisher who accepted nomination as Chairman of the NFFO: '*He's a puppet…you have a U-10m fisherman saying we don't need fish... We have a chair who is an U-10m fisherman and he's saying that… we don't want to give quota to the U-10s... He is definitely a Judas there, accepting the silver coin*' (KI-48).

On political obstacles, three fishers say their small size makes them politically weak: '*the smaller you are, the more vulnerable you are, the less kick you have*' (KI-89); '*rules and regulations are waged against the small people*' (KI-69); '*we have been told… that the amount of fish that the U-10m boats actually catch is so small that is it not even worth bothering with…They see us as such small fry that we can just be brushed aside*' (KI-59). Another fisher explains that unlike the farming industry, the inshore fishing fleet has little lobbying strength: '*there is no lobbying body…for fishermen….all they are is just a bunch of winge-ing fishermen…They* [politicians] *don't need to listen*' (KI-23). A fisher says '*you never see a politician down there anymore. They got bigger fish to fry than those trying to catch less than*

100 votes. They are better off going to a call centre where they could get 700 votes' (KI-68). An IFCA CFO agrees: '*there are very few votes in fishing…so it's not really seen as a big issue*' (KI-105. Another IFCA CFO says '*There is not a political party that represents them in any way, shape or form*' (KI-103). A FA leads says they have no one defending them in the political sphere: '*our government, not one of them would stick their head above the parapet…for us...no one is interested in us and we don't have anyone fighting for us. We are just forgotten*' (KI-61).

It is true there are fishermen's organisations that lobby the government on behalf of the English fishing industry, most notably NFFO and NUTFA. But according to a FA lead, NFFO is predominantly focused on the concerns of the offshore fleet: '*The NFFO supposedly represents the whole of the fleet but they don't, their money comes from the big boats...they care about the people that put the money in which is predominantly foreign investment...It's business at the end of the day…it doesn't come down to morals or the individuals, it's all down to business*' (KI-16). Two fishers say NFFO does not help the inshore fleet: '*we just feel that they wasn't doing nothing for us. They just bypass the under-10m fleet*' (KI-47); '*they are not the inshore fisheries friend*' (KI-64). Another fisher says this is why many U-10m fishers left NFFO: '*We used to be in the NFFO years and then they didn't want the U-10s in there so we left*' (KI-97). As for NUTFA, it is deemed to have no political clout: '*they are so lightweight that they are not worth bothering with*' (KI-74). NUTFA are accused by an IFCA CFO of failing to represent all the inshore fleet since many inshore fishers have not joined it: '*They never have the full voice, so their weakness will be that they are not unified in their approach*' (KI-105). A fisher agrees: '*it has been noted how few members NUTFA actually have on their books. But they're being treated as the voice of the U-10m fishermen*' (KI-44). NUTFA is criticised by a FLAG coordinator for focusing more on white fish vessels than on shellfish vessels: '*Their main interest is more quota, so they are not interested in boats that go for shellfish. They have a specific issue and we don't play a part of that*' (KI-82). NUTFA are also accused by a FA lead of concentrating too much on the southeast to provide a national unified lobby: '*That's why I got out of NUTFA… they have got the worst problems down there in terms of quota, but they are incapable of representing the rest of the country effectively and efficiently*' (KI-100). Another bone of contention with NUTFA is its alliance with Greenpeace. A fisher says: '*that Greenpeace thing worried me. I wouldn't trust tree huggers as far as I could throw them*' (KI-73).

On physical obstacles, the most obvious physical obstacle for the inshore fleet is its small vessel size, which, as two fishers point out, restricts SSFs to areas close to home: '*As a small vessel is not nomadic, it can't travel like a larger boat. A larger boat will chase the fish. It will work out of one port and chase the fish in another, whereas a U-10m fisher will work out of his own port and sleep in his own bed at night. He will just work days and that restricts him to 20 miles within his port*' (KI-64). As another fisher notes, this means some inshore fishers find it hard to diversify by targeting other species: '*The downside of our fleet here is that there isn't any room for diversification. It is a very strong shellfish ground and there isn't a lot else for them to catch…So we are stuck with the crab and lobster really*'

(KI-80). A fisher says small size makes the inshore fishers particularly vulnerable to spatial restrictions (KI-96). An IFCA CFO says with regard to MPAs, offshore vessels '*just move somewhere else...*[but] *an inshore fleet can't do that*' (KI-108). Moreover, when temporary area restrictions are lifted, a fisher declares it is invariably the offshore vessels that benefit: '*As soon as they lift the restrictions on the NTZ, you can bet your bottom dollar that those Spanish boats are going to come in and take everything that you have been saving for the last two or three years. Within weeks*' (KI-24).

An IFCA CFO points out that another physical weakness of the English SSF is its ageing profile: '*in our district...the average age of skippers is high and there doesn't seem to be a lot of second generation fishermen coming through...It's becoming an aged industry*' (KI-109). A FA lead says '*That chap that was talking about how we need to preserve it for the next generation. Well there isn't one. We are the last generation*' (KI-99). A fish merchant agrees: '*I don't think there will be any more fishermen after my generation. What's to attract them?*' (KI-60). An IFCA CFO says young people are deterred by poor prospects: '*Their boats are old. Their people are old*' (KI-110). Wages are low: '*it's so difficult to make a living from it*' (KI-110). Hours of work are unpredictable: a FA lead says '*with a decent job ashore you got set hours. Monday to Friday. They could tell you in six weeks' time when they got days off. Whereas in the sea it's 24/7, especially in the winter as the weather dictates everything. So you could be inshore today, tomorrow, Friday and then even if you have plans over the weekend, if the forecast for Monday and Tuesday is bad, everything has to change and you have to go out to sea. You have to work the weekend*' (KI-1). One fisher says he does not want his children to follow in his footsteps (KI-39), and a FA lead agrees (KI-99). A fisher says young skippers are tempted away from fishing to service the wind energy sector:

> I know some of the younger fishermen who now are skippers on the wind farm boats as they can earn a regular salary, a very good salary of £40,000 every year. They don't want to do it. It's very boring, just running people back and forwards to the wind farms to do maintenance and stuff. They would rather be fishing, but with fishing they don't know if they are going to get an income one month to another. Nine times out of ten they won't get enough to live off from it...this is another example of how the wind farms have been to the detriment of fishing as they are taking the younger skippers away (KI-68).

A fisher explains how this means crews are harder to find—'*Crews have gone to the wind farms*' (KI-71). Moreover, recruits are often quickly disillusioned: '*they get their wages and they get such a shock that they don't come back*' (KI-6).

4.4 Conclusion

This chapter has presented the perceptions of fishers and managers about the threats and obstacles facing the English SSF. We divided those threats and obstacles into external and internal categories. The external threats were subdivided into economic, governmental, and environmental threats; and the internal obstacles were subdivided into psychological, educational, social, and physical obstacles. At first sight, it appears that the external threats are beyond the control of the SSF, whereas the internal obstacles are self-inflicted. There is some truth in this observation, but it is too simplistic because some of the external threats (such as lack of diversification are partly due to SSF behaviour), and some of the internal obstacles (such as educational limitations) are beyond SSF control. The importance of the chapter lies less in the analytical clarity of its distinction between external threats and internal obstacles than in its identification of the extent and depth of the vulnerability felt by the English SSF. It is striking how many of these sources of vulnerability in the English SSF are common to SSF around the world (Jentoft and Midré 2011). The chapter also prepares the ground for the next chapter, in that an account of the way fishers and managers perceive they have responded to their problems (the subject of Chap. 5) makes much more sense following an account of the way they perceive those problems (the subject of Chap. 4).

Chapter 5
Strategies of Resilience in the English SSF

Despite its neglect in policy terms and the absence of special financial assistance, the resilience of the [English] inshore fleet has been remarkable (David Symes)
Many men go fishing all of their lives without knowing that it is not fish they are after (Henry David Thoreau)

5.1 Introduction

This chapter reports respondents' perceptions of the way the English SSF has faced up to the challenges described in Chap. 4. Following the interpretive framework outlined in Chap. 2, we divide these perceptions into three categories: passive resilience; adaptive resilience; and transformative resilience. It is worth noting that not all acts of resilience succeed in overcoming challenges: sometimes a fishers' responses to challenges creates a feedback loop which exacerbates their vulnerability (Holling and Gunderson 2002).

5.2 Passive Resilients

At first sight, passive resilience might appear to be not so much a form of resilience as an admission of defeat or lack of resilience, by contrast to the optimistic responses characteristic of both adaptive and transformative strategies. However, this interpretation paints a one-sided or monochromic picture of passivity: the truth is that many passive resilients hold a very positive self-image of defiant survival despite the odds stacked against them. Accordingly, this chapter interprets passivity in the English inshore fleet as a form of resilience, albeit one that contains both negative and positive mood music.

© The Editor(s) (if applicable) and The Author(s), under exclusive license to Springer Nature Switzerland AG 2021

R. Korda et al., *Resilience in the English Small-Scale Fishery*,
https://doi.org/10.1007/978-3-030-54245-0_5

Henry David Thoreau's observation that "The mass of men lead lives of quiet desperation", captures the negative mood music of passive resilients. For one fisher, the obstacles have taken the joy out of inshore fishing:

> this is me, a fisherman who loves the job, who loved the job. It's not becoming a job, it's becoming a chore. It's stressful…going out there…with bailiffs on your back. With …so many letters coming through your door. It's just a never-ending circle of shit. One time that shit was sweet. It was fun, it was exciting, it was all the things that fishing was. Now it's just a shit of a job...Fishing is not a happy, happy job any more. There are no winners in…it any more (KI-7).

For two fishers, all the excitement has been removed—'*I used to go to sea and you wouldn't know if you would catch one or five boxes. And that was the thing about fishing is that it used to be exciting. It's not any more*' (KI-52); '*It is so debilitating. I can't express the feeling*' (KI-21). Two fishers say they cannot see a future in their sector: '*I am quite gloomy about the future direction of the industry and the people that lead it. They are such muppets*' (KI-39): '*It's just that there is everything against the U-10s…The under tens are a dying fleet*' (KI-10). Another fisher is sad at the break-up of his beloved industry:

> I get so down and depressed about all the skills that I've got and knowledge that I've had handed down to me from my father and grandfather. To see it all being just diluted and broken up by a load of rules that don't seem to make any sense at all. And I'm not alone. This is happening in every British fishing family around the coast. We are watching ourselves being taken apart (KI-21).

Fishers appeared to react to this and their other frustrations by disengaging with formal arrangements such as management and their associations and giving up trying to change things: '*The trouble is that over the years and the decades…fishermen come to all these meetings and nothing changes. And they are disenchanted*' (KI-68). A Fisheries Association (FA) lead describes fishers he works with as being ostrich-like: '*Most of the fishermen just want to bury their heads in the sand and just go fishing*' (KI-61). A fisher says he has learned to put up with whatever is decided by other people: '*I never go to the meetings. I just don't want to go any more as it is a waste of time. It just feels like you are wasting your time. I just think you just need to accept what they come up with and just grin and bear it…most of the times the rules come out, and you just have to accept that*' (KI-89). However, a FA lead says this attitude only makes their situation worse: '*now we are even in a worst situation than we were before because a lot of good people have just walked away and won't get involved anymore and the whole thing is just moved into complete and utter meltdown*' (KI-1).

Nevertheless, a fisher says they keep on fishing. For some, it is because they don't know what else to do: '*a lot of these boats here, they don't make a lot of money out of it…I think it's only 'cause they are old-timers and they have been doing it all their life, so it's like a habit, they don't know how to get out of it*' (KI-8). Reed et al. (2013, p. 65) characterise them as survivalists: "They were 'survivors' in a number of senses, not least because they have remained in fishing despite its decline…They tended to run their businesses in 'survival' mode, keeping overheads and risks low, whilst not looking to reinvest capital in their business".

According to an Inshore Fisheries Conservation Authority (IFCA) Chief Fisheries Officer (CFO), passive resilients have a traditional mind-set rather than an opportunistic mind-set: '*I think that they often aren't sufficiently business-orientated. They operate in a way which is very traditional*' (KI-111). A fisher agrees: '*They want to fish, they want to sell their product and then they want to go home. And we have not had many entrepreneurs who have seen the opportunity...there is a traditional mind-set in terms of they know their business...so they haven't quite woken up to the other opportunities with might exist. So there are strengths but also a bit of weakness as they are stuck in their ways*' (KI-83). According to a critical IFCA CFO, passive inshore skippers continue to maintain a cottage industry perspective despite facing a global market:

> So it's marrying that global perspective, with a very, very, very local cottage industry perspective of "this is my boat, this is the fuel it burns, this is my business model, it's the one that my grandfather had and my father had, this is the one that I use", and those two things don't marry up…Rather than recognising that there are bigger forces at play and no matter how much you fish using your current business model, you are never going to make it viable, you need to adapt and do things differently (KI-106).

For example, in response to the possibility of gaining an accreditation status through a scheme such as that of the Marine Stewardship Council (MSC), the passive resilient's response is that '*no one will pay a premium for line-caught fish or sustainable MSC*' (KI-88). Likewise, in relation to the quota leasing system, some fishers see it as too risky: '*we…leased in a tonne of cod. Then it was down to the weather, whether we caught it. If we didn't catch it by the 31 December then we lost it. Which we done two years ago…that was £700 down the drain*' (KI-58). An IFCA CFO says this perspective makes them turn their backs on change: '*they're not really embracing the changes*' (KI 110). Some passive resilients wish to turn the clock back: for example, a FA lead wants a ban on dredging: '*Dredges aren't no good for anything…if they went back to old-fashioned methods, the world would be a better place…and everyone would still get fed*' (KI-2). A critical IFCA CFO attributes this conservatism to inshore fishers' isolated working environment: '*a failure to move with the times...is symptomatic of a people who work…away from the land and you are disconnected a lot of the time and you don't tap into the normal feeds that everyone else does on a daily basis…You are slightly isolated, so when things change it comes as a big shock*' (KI-106). A Fisheries Local Action Group (FLAG) coordinator says some fishers are not keen to seek alternative livelihoods outside of the fishing sector because of the vocational nature of the sector: '*It's a lifestyle not a job*' (KI-14).

By contrast to the gloom-laden negativity of the above passive resilients, other passive resilients are more optimistic and positive about their prospects. Some passive resilients explain that they continue to pursue existing fishing practices because of a commitment to sustainable techniques: '*I think if you look at it as one sub-sector on its own working a small section of sea, it is very sustainable as they don't take a lot and what they take out is not large-scale in any way*' (KI-87). A FA lead says passive resilients are proud of their light marine footprints—'*We pride ourselves on the fact that we are sustainable...What they don't realise is that small boats like ours are the best conservation you can get. The herring fishing here has been going on for over 500 years, there's nothing more sustainable than that. The lobster fishing here has been going on for hundreds of years, we don't catch that much*' (KI-3). A fisher says he is committed to upholding traditional methods: '*I use traditional fishing methods, static nets on the sea bed. It's very passive and it means that fish need to swim into the net. I don't drag it so I don't spoil anything. I also use a hand-line. So it's a very passive fishing method*' (KI-22). An IFCA CFO says they should be praised: '*The sustainable fishermen should be celebrated*' (KI-106). An FA lead says their traditional practices ensure there is fish for everyone: '*if you work it like we always worked it, then there is something in it for everybody*' (KI-2). A fisher claims their operations enhance local wildlife: '*That's like our mussels. We have been doing mussels for eight years. And the amount of birds that we feed and more or less keep. There are 300 oyster catchers every day. All natural. They know us. These conservationists don't really know what they are doing, as if it weren't for fishermen, half these birds won't even be here*' (KI-75). An IFCA CFO describes passive resilients as subsistence fishers: "*there are some subsistence fishermen... they go out for the day get a basket of fish, buy some fuel for the boat and go and repeat and if they have a good day they make a profit and if they have a bad day they don't. The good outweighs the bad*" (KI-106).

Some of these viable fisheries for the U-10m fleet are geographically very specific. For example, two fishers say they are lucky to have ample access to non-quota fishing opportunities, such as shellfish and red mullet, so they do not have to change their operations (KI-72; KI-29). Another fisher says he has easy access to the sea: '*It's also very quick to get out to sea it's right on your door step so no rivers to cross*' (KI-13). A FA lead says he is based in a sheltered bay which means that when other target grounds are inaccessible because of bad weather, he can continue to prosecute his local grounds (KI-12). A fisher reports that in his area there are no threats from nomadic fleets (KI-80). A fisher says that no-trawl areas protect their fishing grounds from more destructive vessels (KI-44). Another fisher says such closed areas have replenished stocks: '*there has been a restock of the sea*' (KI-68). A FA lead says another geographical advantage for some English SSFs is a lack of sensitive ecosystems: '*We are fortunate as we don't have any eel grass here and we don't have much contact with the environment groups*' (KI-42).

However, above all, what keeps many passive resilients going is the personal pride they draw from their profession. As one fisher says, their self-image is one of heroic seafarers braving the elements in treacherous conditions to bring home nutritious food for their families and communities:

> We are a food-producing sector full of what I consider to be very brave individuals who pit themselves against the elements and all that the weather can throw at them and come up with a...wonderful source of food...You have to have a lot of guile, a lot of skill, a lot of bravery and a lot of intelligence to survive the ocean and produce a catch of fish that will cover all the costs of the day and the business and provide for your family...to me, fishing is more than just an industry. My family have given one of our own to it. The bottom line is

> that this is a dangerous job, it comes with a price... I think of the blood, sweat and tears that have gone into it from our side (KI-21).

An IFCA CFO expresses his admiration for U-10m fishers' courage:

> I think their sheer courage and bravery and determination in getting out there, in what I think is still the most risky occupation bar maybe one or two that anyone can do anywhere, just fills me with admiration all the time. Their resilience, their ability to overcome setbacks and really just to keep going in the face often of…natural disaster, in the face of the sort of pressures…which come from government…stepping on the toes of the industry (KI-111).

A fisher and a FA lead say fishing is more than a job: '*It's a vocational thing*' (KI-39); '*it ain't a job, it is a way of life*' (KI-72). It's a love affair which keeps them going: '*It's getting harder and harder. I honestly think is it really worth staying in the industry the way it's going? I love the job, that's the only thing that keep me here*' (KI-19). One fisher describes it as an addiction: '*Fishing is a very difficult thing actually to give up. It's like a drug in many ways, because it's like an adventure. It's a boy's adventure and once you are committed to it and your life is given over to it, everything you do revolves around it…because you come to shore and your next moments are that you are looking at the weather planning when you can get back out there again*' (KI-21). Another fisher says what drives them is not money but 'buzz' and 'street cred':

> it isn't money that drives us. It's the buzz. Yer, we are there for money but we are also there for that buzz. And if you don't like that buzz then you won't go out fishing. As there are some times when you are dog-tired and fuck-all is going right and you can't give in because you have dragged your crew out on a Saturday and you have to prove yourself. It's about street cred and if you ain't got that drive in you, you will never make a skipper…I wouldn't have me life another way (KI-78).

Another fisher says fishing makes him feel like a rock star: '*It's the rock style lifestyle ain't it? Getting up at two in the morning, six days a week. I love it as a job*' (KI-10). An IFCA CFO says fishers are born not made—'*I think you're born to it, not made to do it*' (KI-111)—and two fishers say there is no question of them doing anything else however bad it gets: '*I don't think I would ever do anything else really. My mother always told me it was a waste of a grammar school education, going to sea. Perhaps she's right but I wouldn't change*' (KI-73); '*We don't know how to do anything else*' (KI-59).

Two FA leads say they are driven to ensure that their fishing communities survive beyond their retirement: '*I'm in this job because I'm passionate about the fishing industry and I want to see a fishing industry when I retire in 30 years' time*' (KI-12); '*I can trace my family back to 1720 and I think out of the 30 members of our Fishing Association, there are only two that can't…And I've got a son and a nephew who would like to get involved in fishing. So even if they don't do it full-time, it's our heritage and we want to pass it on*' (KI-3). A FLAG coordinator says the same: '*I feel passionately about the sea. I come from a long history of seafarers. My family were seafarers for hundreds of years and I want that to carry on in my family. And unless I stand up for it, it aren't going to happen*' (KI-66). An IFCA CFO says fishers were driven by a strong sense of identification with their fishing community: '*People are very passionate about where they live here and we would defend it to the last person if it was attacked. Everyone would here. Very strong indeed...We have so many links. A lot of people are related anyway over here. The population is on two different islands through marriage and through blood*' (KI-107).

It may be that another factor explaining the persistence of subsistence fishing is the psychic gratification of being admired by tourists and sightseers – an experience an IFCA CFO describes as follows:

> It has a big effect on tourism. You see the tourism posters for XX and it's all based around pretty fishing boats, pulled up on [beaches in] pretty fishing ports. And people do like to come and watch it and they do like to come and see what's happening with the industry. When the boats land up here on the beach and are winched up the beach and unloaded, there are a lot of people who are watching... you've only got to have a boat refitting alongside the wall here and there will be holiday makers hanging over the wall here, talking to the fishermen and asking them what they're doing...[It is] the theatre of the fish market more than anything. You get a lot of film crews coming down and requests from people to try and see it. They now do fish market tours which have made it onto the Observer list of top ten things to do in the area...it's exciting. It is different and it's not something that everyone is used to (KI-106).

One fisher expresses optimism among passive resilients that there are some young people who are interested in joining the inshore fleet: '*There are certainly more youngsters around than there used to be...ten years ago*' (KI-27). An IFCA CFO agreed: '*Youngsters do want to go to sea*' (KI-105).

5.3 Adaptive Resilients

The difference between adaptive resilience and passive resilience is that the adaptive fisher does not passively make the best of a bad job, but actively searches for new ways in which to take advantage of the situation which confronts him/her. Unlike passive resilients, adaptive resilients are willing to radically change the way they work, and to adopt innovative strategies to exploit the opportunities which present themselves. Adaptive resilients respond to changing circumstances by changing their behaviour and their mind-sets to embrace the new realities. Unlike transformative resilients, adaptive resilients do not fight the new realities, but constantly adjust to them. The ways in which adaptive resilients adjust to changing circumstances are many and varied, though not all adaptive strategies are sustainable, and this section is divided into two parts: unsustainable and sustainable adaptive strategies. There are two unsustainable adaptive strategies—illegal practices; and solo working. There are five sustainable adaptive strategies operational flexibility; reducing overheads; quota tactics; marketing strategies; and job diversification. Some of the responses are strategic—i.e. long-term shifts based on forward planning and serious reflection requiring investment—other responses are tactical—i.e. short-term opportunist shifts, requiring quick adjustments to fast-changing circumstances.

The first unsustainable adaptive strategy is illegal activity. A desperate form of tactical adaptation to reduced quota is to breach the regulations. For example, a fisher explains that faced with the prospect of not being able to pay his crew's wages, he ignores quota restrictions: '*I know if I get caught, I'll get hung out to dry. So what do you do? Do you chance it? Sometimes yes, 'cause you have to. If you don't bring the next two or three boxes up, what happens then? Your crew get nay wages. What happens then? They don't come back*' (KI-7). A fish market owner admits '*I have broken laws, 'cause of the stupid regulations that they have put us under...when you are running a boat, and you need 100 boxes of fish to give the crew a wage and they turn round to you and say you can only catch 50 boxes what do you do? Do you lose your boat, do you tie up and go into debt?*' (KI-79). Likewise, a fisher (KI-21) says '*I've fished illegally as there was no other way of fishing to make my business work*'. The second unsustainable adaptive strategy is solo working. Some fishers reported that when faced with financially unviable operations, they move to dangerous lone working arrangements: '*I fish alone most of the time. This has changed because of the lack of quota to pay crew*' (KI-71). This was deemed to be a health and safety issue: '*it's quite hair-raising when you're on your own I will say*' (KI-18).

The first sustainable adaptive strategy is operational flexibility. This is the most common form of adaptive resilience displayed by inshore fishers and was said to be embedded in the very nature of an inshore fisher (KI-106). A fisher says flexibility is ultimately key to their survival: '*it has always been an industry that has to change with the times, and some people don't like change, but I think that if you are going to succeed you have to recognise that some stocks are no longer viable*' (KI-40). An IFCA CFO notes that many fishers have a pronounced capacity for innovation and adaptability: '*they are very flexible, so they have developed the ability to be pretty multifunctional and to capitalize on opportunities*' (KI-107). For example, inshore fishers adapt to the seasons: '*you can jump from one fishery to another and in an ideal world you would do one fishery in the summer and another fishery in the winter. You would work the seasons*'. Such tactical shifts enable inshore fishers to work all year round. A fisher explains that offshore vessels chase the fish, but inshore boats cannot do so, so they change their target fish: '*One day they can be working the whelks or the lobsters and crabs and then they can change when the cod show up*' (KI-64). Other shifts are driven by quota regulations: two fishers give an example of strategic operational flexibility in the permanent shift from fin-fish to shellfish: '*there are several who had trawlers and changed over to pots, just because of the restrictions the MMO was putting on us*' (KI-98); '*When I first went to sea, you wouldn't have made a living out of crab. Now it's our mainstay*' (KI-73). An IFCA CFO says; '*These guys...show great adaptability...from being predominantly a trawling and white fish fishing operation to shifting their emphases to shellfish*' (KI-111). This is a shift which requires a significant amount of investment in new gear.

One fisher says he has made this shift for moral as well as economic reasons—to avoid illegality:

> For the last ten or twelve year, I've been at the lobster pots...the previous five years to that I owned my own trawler. But due to the quota restrictions by the MMO or Defra as it was then, I personally felt that it was better to get out of it as I thought it was going to lead down

> the line of having to go illegal. And I wasn't really prepared to do that as I try to lead an honest life. So I thought going to the lobster pots was a better venture as we could do things legally (KI-97).

However, an IFCA CFO points out that this adaptive strategy may reduce fishers' options in the future: '*The U-10m fleet is quite strong in its ability to change methodologies on the hoof. But it's also a bit of an Achilles heel as…they have been more and more painted into a corner*' (KI-104). As another IFCA CFO explains, this is because the government's return to capped licence arrangements means that inshore fishers who move from fin-fish to shell-fish risk losing their licences to return to fin-fish, and this would leave them exclusively reliant on shell-fish: '*A few years back, the UK government reviewed the U-10m licensing system and many of the vessels in our district had their licenses capped, which meant that they couldn't access other species or lost any sort of level of access to species which meant basically the majority are now almost exclusively reliant on shell-fisheries*' (KI-108). A third IFCA CFO notes that the shift from fin-fish to shell-fish has put greater pressure on shell-fish stocks, which will become permanent if fin-fish licences are forfeited: '*as people have fallen out of fin fish licenses and got shell fish licenses, there is an awful lot more effort on smaller grounds*' (KI-105).

A FA lead gives another example of operational flexibility—inshore fishers' willingness to experiment with new fisheries techniques and areas: '*Every year you might just try something a little bit different and you might just find it works just a little bit better than how you used to do it. You keep testing your working practices*' (KI-100). This is tactical adaptation. Another FA lead reports a more strategic move by his FA investing in ice plants and storage facilities on shore to hold catches until market conditions logistics improved: '*We are now looking at a new project of over a million pounds of revamping the fisheries quay. The ice machines will be replaced and new buildings brought up-to-date…and a platform around the other side for extra landing*' (KI-26). Taking advantage of the FLAG scheme is another strategic example of operational flexibility. A fisher explains how grant money sourced by their local FA by engaging with the FLAG program has funded improvements in infrastructure to provide fishers with easier access to the sea (KI-37). A fish merchant describes how their local council helped fishers win a FLAG award (KI-67). An IFCA CFO refers to FLAG schemes designed to actively recruit youngsters to enter the industry (KI-107). A fisher explains one such scheme:

> The FLAG is trying to help with that with its program, Net to Plate, where it is going into the schools. The fishermen themselves would say most kids these days would never even think about going out into the boats as it's not even worth it and if they are not born into it, they won't do it. But the idea is at least to try and raise some interest…and put some future proofing into the sector (KI-37).

Another fisher explains that a FLAG grant enabled him to set up a processing business for crab meat which is all sold locally: '*you gotta try and make the best of what we are catching…we process all our own crab meat so we don't send any crab meat away…we got a small processor out the back of the house…we shift all our crab meat locally…*[to] *local hotels and restaurants and local people…and to people coming on holiday*' (KI-37). A fisher says FLAG funds have been used to support fish festivals: '*you had the crab and lobster festivals. That got a grant. So that is promoting the fishing industry*' (KI-65). These are strategic adaptive responses. Van de Walle et al. (2015) testify to the value of the FLAG scheme in Brittany in facilitating the adaptive resilience strategy of the SSF at Pays d'Auray. Phillipson and Symes (2015, p. 352) refer to the FLAG scheme as a "bottom-up initiative set within a command and control system of governance".

The second sustainable adaptive strategy is reducing overheads. One way of doing so is by collaborating with family members. A FA lead explains how her family shares jobs out amongst themselves that they might otherwise have to outsource: '*my husband does all the wheeling and dealing and I do the paper work*' (KI-1). A fisher says his family divide into fishers and processors: '*if you can find a little place where you can pick the crab meat…So you can supply the restaurants directly…my wife and my mum…do that in the morning*' (KI-36). Another fisher says his family members include solicitors, a resource they are able to utilise free of charge: '*My daughter and nephew are…lawyers, making sure that* [lawyers] *are not shafting us…they are working on this for free…My daughter looked at* [the evidence underpinning a proposed marine protected area (MPA)] *too. The problem was that you had to challenge it with evidence. So she tore their evidence to bits and that's the only reason why* [MPA] *still hasn't been approved out of the other 27*' (KI-62). The same family was also able to draw on the office management skills held by their other daughter to put forward suggestions on how the IFCA could organise themselves better: '*My daughter suggested that it could be done differently. She was an office manager at some point and they would write up the cases for prosecution and you have to know what is going on…Her job was organising people really and the IFCA don't have a clue*' (KI-62).

Note that these strategies all involve women, and this raises the question of the place of women in the English SSF. The number of women respondents in this study is seven—four fishery association leads (of whom two are active fishers); two merchant owners/managers; and a FLAG lead. Their responses reflect the male-dominated culture of the English SSF, with the fisher women referring to themselves as fishermen, and explaining how they got started in their roles off the backs off their husbands: '*ever since I got married, XX and I go to sea*' (KI-1); '*My husband was a fisherman back when we got together 11 years ago. And he had just come ashore to work for this company and I was looking for a new job. He recommended me here…So it will be 11 years in April that I have worked here*' (KI-79). One of the fisher women spoke with great pride about how she regularly out-fishes the men: '*People think I fish with Trevor but I fish in my own right. Whatever the chaps catch, if you can rake a tonne of cockles, I can rake a tonne and a half*' (KI-1). However, for the most part, the work of women in the English SSF is onshore, supporting rather than competing with the fishing activities of their male counterparts. SSF discourse reflects this secondary position of women. For example, a fisher says '*I have five daughters so I don't have any sons to go into fishing, but most fishermen have sons and they will want them to carry on their work*' (KI-52). A fisher says of a fellow fisher: '*he has got grandsons coming up and I would love to see them down there*' (KI-64). Even management is talked about in male terms: '*they are mostly young lads*' (KI-89), although one woman in the Department for Environment, Food and Rural Affairs (Defra) has been quite widely publicised as she manages shellfish legislation. In our view, this largely ancillary role for women in the English SSF is a missed opportunity to take full advantage of the qualities and skills which women would bring to the inshore sector if they were given more encouragement to become involved in the fishery.

Another respondent explains fishers are able to cut costs by sharing skills across their community: '*Being able to help each other, because not everyone has the same strengths. So I'm good with engines and the next boat is useless with engines but he's good with electrics, so yer, we help each other and it keeps the costs down ... There is another fishing cooperative in XX and we do try to all work together...to keep the costs down*' (KI-6). A FA lead says another method of reducing overheads is obtaining cheaper fuel through FAs: '*The association used to buy the fuel in so if you were in it then you got it a little cheaper*' (KI-99). A fisher explains how he brings costs down by targeting species which have low cost gear: '*a whelk pot is fantastic- it's the only thing that will catch a whelk. It's a light weight pot, it don't cost much to make. They are cheap as the trawlers throw them around out there and smash them up so they need to be cheap*' (KI-54).

The third sustainable adaptive strategy is quota tactics. Although inshore fishers generally denounce the quota system as a disaster for their sector, adaptive resilients use the system skillfully to their advantage. For example, one fisher sees the leasing of quota as an opportunity: '*I have got an option...rather than throw* [fish] *away...I can lease it from...someone who might have it and isn't going to catch it...it means that that fish doesn't have to be thrown away*' (KI-21). A fish merchant quota holder defends his leasing out of quota as enabling adaptive fishers to continue fishing: '*If they can't go fishing, then we can't have a fish market. So I am not doing it for altruistic reasons, I'm doing it to try and keep everyone in business...Personally we don't run quota to make money. We run it for access to the fish...In fact, quite often, 'cause I feel mean doing it, we end up...subsidising them. But if I don't give quota to these fishermen...the whole of the fleet will stop*' (KI-51).

The fourth sustainable adaptive strategy is about marketing. Respondents explain how they adapt their marketing strategies to suit the market opportunities available to them. For example, a shellfisher says he exports his produce abroad to take advantage of strong prices offered on the continent: '*Chances are, if you go to a restaurant in Paris, you will be eating Bridlington lobster, whereas if you go to a restaurant in Leeds, you will be eating Canadian lobster 'cause it's cheaper*' (KI-82). A finfisher says he benefits from close proximity to international markets: '*France is only 32 miles from here. The fish goes down to Dover and we are only 18 miles from Dover and that is where the ferry goes from. So the lorry goes round and picks up the fish and runs over to France and can be fresh on the market in the morning. So we are lucky here*' (KI-63). Another fisher says he benefits from his close proximity to London: '*a lot of it gets exported straight into the London restaurants*' (KI-83). Another fisher says new electronic auctions help to maintain good prices for high quality inshore landings:

> An electronic market…[ensures] that the buyers are not actually sat next to each other. Because whilst buyers are in competition with each other, as the fish comes in…you can negotiate with the guy sitting next to you to bring the price right down…if I take this box you can take the next one and I won't haggle the price up for you. So then both you and your neighbour can pay bottom prices for top end fish…So it's not a fair bidding system. Now with an electronic market like they have at XX, you don't know what the other is bidding, so it's like a silent auction…It works in our favor as we land high quality fish compared to the deep sea trawlers so we get the higher prices (KI-22).

A FLAG coordinator has adopted an alternative strategy—selling his produce directly to customers: '*selling fish more locally, like directly through the local restaurants rather than it having to go to a market a fair distance away*' (KI-5). (KI-104). In some locations, a fisher says FAs are the driving force, setting up direct selling schemes for their communities (KI-37). Other fishers collaborate to transport their fish to more lucrative markets: '*We share a vehicle that takes the fish down to XX … we try to get the costs as low as possible but get the best prices back for the goods that they are selling*' (KI-6). Another common strategy is to create a niche market for inshore fish: '*I think there are some people who are very good businessmen and they work a lot on developing unique selling points for their products and have developed quite a bespoke market*' (KI-108); '*Some of them are doing very, very well. The lot I like to call the go-getters. That is the people who are very proactive and either will find niche markets or will do their own selling*' (KI-104). For example, they emphasise the freshness and quality of their products: '*The inshore sector brings in their fish daily straight to the market…in our case, we fish it today and people eat it in the evening and that is where the strengths of the U-10s lie…that we are harvesting daily the best quality seafood that you can buy. It's alive and it's fresh*' (KI-21); '*what you have to try and do is make sure that what you are selling is good. So the prawns are good and the fish is clean and it's a good size*' (KI-4). A FA lead says such bespoke marketing means low quantities may be more remunerative than large quantities, so fishers can afford to land less as the value of their catch has been increased: '*They have to use less efficient gear now, with big square mesh panels in larger mesh cod ends so they are not catching the bulk. But what they are catching is of a better quality…they are landing less than half of what they were landing 20 years ago but they are making more on it…the quality has increased but the quantity has decreased*' (KI-100).

An IFCA CFO says part of this niche marketing strategy by the inshore fleet is to stress its superior environmental credentials: '*the Cornish hand liners, they fish using hurdy gurdies and catch in a sustainable way. And they market their fish, they have little labels that say "Hand-lined caught in Cornwall". And you can add a premium for that*' (KI-107). A fisher reports this had recently been recognised by a European ruling: '*Article 17 Europe ruling says that those fishing inshore are more sustainable*' (KI-44). Another fisher attributes their environmental superiority to employing non-damaging techniques such as potting, which is '*a very non-destructive way of fishing. The pots float to the bottom and act as a feeding station for all sorts of wildlife, not just the commercial catch, and that has helped the success of the fishery. It is almost like farming. Not like trawling where you are just wiping everything out the way*' (KI-80). Another fisher says long-lining '*is the most environmentally friendly, conservation-minded way of fishing there is…Very little discarding. What we catch we keep. We are not like a trawler who catches 15 boxes of fish and goes through them and picks the big ones out and chucks the small ones back*' (KI-72). This demonstrates that adaptive resilients do not see environmental regulations as their enemy. Four fishers say '*Most fishermen, even though we are portrayed that we want to go out and grab everything, most fishermen know they need to look after the sea. We need to protect our future*' (KI-74); '*It's great to catch lots but if you don't look after the future then it is going to run out and we want this to carry on for a long, long time*' (KI-80); '*The real fishermen are in it for the long haul*' (KI-91); '*There is total agreement with minimal landing sizes; you won't find any arguments from fishermen*' (KI-23); '*We are not catching everything we can. We are looking after our stocks as we don't want to shoot ourselves in the foot. We haven't got another patch of water we can move into. We have to look after it here*' (KI-35). Another fisher says the inshore sector are the true conservationists: '*You see there is no one more conservation-minded that an inshore fisherman…Working as inshore fishermen, conservation is paramount in your mind. I am not alone in this thinking. We all want our tomorrow*' (KI-66).

Of course, as we saw earlier, passive resilients also claim a light environmental footprint. Where adaptive resilients differ from passive resilients is that passive resilients claim their environmental credentials are longstanding in traditional methods of fishing, whereas adaptive resilients admit their environmental stewardship mind-set is a recent phenomenon. A fisher says '*More recently, the people are much more aware that they have to look after what they got so they are not so blasé about what they are landing and landing stuff they shouldn't be landing*' (KI-37). An IFCA CFO says this reflects the greater environmental awareness that has occurred in society as a whole: '*The increased awareness of environmental issues in society…a lot of the industry…readily embraces. They realise that conservation of stocks clearly is a major aspiration to ensure the long-term success of fishing activity*' (KI-111). A market owner claims this change of heart has given fishers more opportunities: '*The opportunities are endless now that fishermen have accepted that there is a need to conserve fish stocks and act sustainably, not only for themselves but for generations. So you no longer go to sea to catch the most fish possible*' (KI-20). A fisher explains that peer pressure kept them up to the environmental mark: '*I don't know anyone who lands anything undersized and sells it under the counter. You wouldn't want to as it's embarrassing to be found out and everybody would look at you and think – you're catching my next year's lobsters*' (KI-22).

An IFCA CFO claims that younger recruits are more highly educated in environmental conservation than older fishers, and therefore more likely to be willing to take these strategic steps:

> one of the good things that has happened in recent years is that fishermen coming in as new fishermen, basically anyone under the age of 30, has had an education in which the environment is part of the curriculum…If you look at my cohort, people that are in their 50s, 60s, 70s, they didn't have the benefit of that…So…the younger fisherman...is very aware of the

> marine environment surrounding him and he is also very aware that he has to carry his activities out in a manner that is not damaging to the marine environment (KI-104).

However, another fisher points out that the quota allocation system based on track records works against environmental stewardship:

> If I went to go and prosecute that fishery, I would have to throw fish over the side as I don't have enough quota for it and therefore as a fisherman…I have taken a moral stand to say I will not go and throw fish over the side and pollute a fishing ground, I'm going to keep the nets in store and earn very little during a period when I could go and catch fish and make money. But…now they are going to penalise me because I…haven't got a track record in it…How can we carry on as businessmen, trying to make the right decision, the honourable decision, the moral decision? Instead, we are being made to make crass decisions (KI-21).

The fifth sustainable adaptive strategy is job diversification, whereby fishers diversify into non-fishing activities to supplement their fisheries income. Some of these job diversification choices are strategic, such as parallel careers; others are tactical, such as temporary work. An IFCA CFO explains how inshore fishing often fails to provide consistent work: '*we are shellfish-centred and…you can catch them for seven, eight months of the year and that's it. You don't catch them in the winter and weather conditions are so bad here throughout the winter…It's definitely not a full-time occupation; you've got to have another string to your bow*' (KI-106). Another IFCA CFO says very few members of the inshore fleet are full-time fishers: '*fishermen nowadays are not just fishermen. It is part of a business portfolio. There are very few single trick ponies any more. Most of them have several irons in the fire of which fishing is just one*' (KI-105). For example, a FA lead has diversified into agriculture: '*my husband and my son fish through the winter and sometimes the summer. But he is also a farm contractor. You see we have to do two sorts of jobs to keep the income coming in*' (KI-1). A fisher takes advantage of local tourism opportunities: '*so I* [fish] *during the winter then in the summer take passengers out as in the summer the crab catch drops off and so you have to find different types of business*' (KI-10). An IFCA CEO says some fishers take on survey work for management bodies: '*we hired their boats to go and survey those areas*' (KI-111). These are strategic choices. Housing maintenance provides supplementary income for another fisher: '*Some of us do a bit of painting and decorating in the winter*' (KI-11). Another fisher uses his engineering skills: '*over the winter I am a tutor and a software engineer. You have a balanced approach to it - you need to have many strings to your bow. If the weather is nice and you can go out fishing you do. If it isn't, you do your other job. You have to be adaptable*' (KI-35). These are tactical choices.

This analysis has shown that fishers who constantly adapt to their circumstances whether by strategic or tactical choices (or both) display considerable resilience in the face of mounting threats to their survival. A fish merchant expresses optimism about the capacity of the U-10m sector to adapt to the future: '*I don't believe these doom-watch people. We built a U-10 boat launched in February 2015 ... Now I wouldn't be doing that and investing the thick end of three quarters of a million pounds if I didn't think that the industry has a future. It's just a different future and you need to adapt to it*' (KI-51). However, this adaptive strategy comes at a high price: constant adaptation is very stressful, time-consuming and costly. No wonder many adaptive resilients have turned to a transformative resilience strategy in the hope of constructing a less stressful future for the inshore fleet.

5.4 Transformative Resilients

In contrast to both passive resilience, which entails resignedly putting up with increasingly difficult circumstances, and adaptive resilience, which entails actively seeking ways in which to get round those circumstances and exploit new opportunities wherever they may be found, transformative resilience entails a refusal to accept those circumstances and instead pro-actively seeks to change them. In other words, transformative resilience is an avowal not to work within the existing system but to challenge it and either overturn it or at least modify it. We can distinguish between two kinds of transformative resilience: radical and moderate. Radical transformative resilience seeks to overturn the system; moderate transformative resilience seeks to modify the system. Underlying both kinds of transformative resilience lies a further distinction—between transformative actions and transformative aspirations. Respondents provided examples of both transformative actions and transformative aspirations. We have concentrated more on the former, on grounds that actions speak louder than words. However, we recognise the importance of aspirations since without aspirations there would be no actions, and so in what follows we mention the most important aspirations that emerge from the interview transcripts.

Radical transformative resilients seek to overturn the current system which governs the English inshore fleet. One reason why respondents embrace radical transformative resilience is because their efforts at moderate transformative resilience have failed. For example, a fisher explains his frustration at the refusal of the management bodies to adopt sensible quid pro quo deals:

> we as fishermen…offered [seasonal closure of spawning grounds] as a solution to improve fish stocks, thinking that by offering that up as a fishing industry we would have some benefit. But what they've done is took that, closed that and then still killed us on quotas. So we didn't benefit from it at all. In fact, it made life even more awkward for us. So as an industry we feel as if we have been cheated, offering up ideas which could be grasped by them and with no benefit to us (KI-21).

There are five forms of radical transformative resilience: self-governance; quota management; fishers' organisations; 'bolshie' fishers; and management attitudes. On self-governance, the most radical form of self-governance advocated by respondents is Brexit – the UK campaign to leave the EU, and thereby the CFP. One fisher sees quitting the EU as a means of reclaiming national power over fisheries: '*We asked time and time again as British fishermen, can we not get out of the Common Fisheries Policy and bring fishing back into the hands of the British fishermen? Back into the hands of the British people*' (KI-21). However, a fish merchant claims the CFP's relative stability principle that is used to allocate quota to Member states, safeguards the U-10m sector: '*Relative stability is the framework... that actually protects us. If we didn't have relative stability, because of the numpties they've got in Whitehall and Newcastle, we would actually have no quota left whatsoever, as there are far better players at the game of quota management elsewhere in Europe than we will ever be*' (KI-51).

Another form of self-governance is decentralised governance within England, which is seen by one fisher as a means of creating bespoke, locally relevant management:

> The biggest problem is too many people interfering at a national level…when issues need to be dealt with at a local level. I think that each area has its own individual problems. You can't just lump us all in together and say well this is what we are going to do as it's good for this bit of the country. That ends up being disastrous for other parts of the country….That's what gets most people's backs up… because it does affect the smaller ports, the smaller boats…quite badly (KI-37).

A fishmarket owner expresses this demand in terms of a bottom-up system: '*we want far less of a top-down approach. We want much more of a bottom-up approach, more regionalised*' (KI-20). A FA lead says '*We do not want to be told…what we can or can't do. We want to look after our own fishery*' (KI-38). A fisher claims '*We're the best stock managers because we want the fish next year and the year after*' (KI-13). A fisher describes how they set self-regulations that work well:

> As a fishermen's association we can make our own by-laws. We are the only ones in the country that can. Because we look after our own affairs they use us as an example to other places actually. At the moment we have control over our own sort of thing…they have left us pretty much alone really because we have come up with solutions to a lot of things off our own back and solutions to things that they hadn't yet started on, but we knew they would do (KI-37).

An FA lead describes another example of successful self-regulation: '*We've brought in a whole load of voluntary measures for v-notching and extended carapace landing sizes for lobsters…Closed seasons in the winter. So we are doing our bit…We have put forward, I think it's 11 measures at the moment which are voluntary…like banning the sand eel fishery to help the sea bird colony*' (KI-38). A fisher says '*the fishermen put a marine nature reserve round here*' (KI-10). Another fisher says '*We proposed our own* [environmental regulations]. *We were quite groundbreaking and jumped the gun...So we were...left alone as we gave them more that they wanted so they were happy for us to work it all out*' (KI-36).

On quota management, many respondents call for a radical transformation of the quota system. One FA lead says quota leasing rates should be related to the sustainability of the fisher's method of fishing: '*If the type of method you use for fishing is like a dust pan and brush method and you don't do any damage to the sea floor, then you would get a better rate of leasing than if you use a chain matt*' (KI-63). Another fisher suggests removing all quota restrictions from 'genuine' under-10m vessels (i.e. not including the super U-10s): '*the best way to solve it is to take the genuine U-10s out of quota altogether and run the inshore vessels on effort limits*' (KI-43). An idea that gained more support was to renationalise all quota: '*The government should take all the fish back from the POs, as…the U-10m fleet only get 2 or 3% of the total TAC and yet there are more U-10m boats than there have ever been big boats*' (KI-72); '*the quota should be a national resource and should be owned by the government and the people of the country*' (KI-20). This would allow quota to be redistributed more equitably between the offshore and the inshore sectors: '*you need to get all the quota back in one pot and then have it shared out again in a fair way*' (KI-67).

Another way of increasing the inshore fleet's quota would be to decentralise power over quota management. A fisher explains how they set out to establish an inshore fisher-run producers' organisation (PO) to operate at national level and obtain the same benefits as the industrial POs: '*the only way I can see proper representation coming from the U-10m fleet, is by forming an inshore PO, and that's what we're trying at the moment. If we get an inshore PO then…very quickly that PO will be stronger than any other PO - by definition* [because]*...70-odd % of the fisherman will be in it so that way everybody feels empowered...the only way things are going to change…*[is]*…being strong enough to actually combat those who have got the money at the moment*' (KI-16). A fishmarket owner sees an inshore PO as a potential game-changer for the U-10m fleet: '*There is talk of them forming a PO, which would be a very good thing as that would give them a bit of political clout. I would regard it as a matter of urgency and I think it is the one and only chance to protect what I would call the small boat fleet. It needs protection because in the end it is likely that its quotas will get swallowed up by the larger vessels*' (KI-20). In 2017, the Coastal Producer Organisation was recognised by the government and incorporated, and it now has 250 small-scale fishers as members, which gives the English SSF a seat at the UK's fisheries policy table and a collective voice to stand up to the dominant industrial fisheries POs. However, the quota allocation crisis remains, because the Coastal PO needs to be given sufficient quota, but, so far, Defra has failed to give quota to the Coastal PO. The need for quota also undermined the attempt by two Kent communities in 2014 to establish a Catch Quota Management Scheme to manage their own quota (KI-63).

On fishers' organisations, the most common means by which respondents perceive that radical reforms could be articulated and moved forward is by fishers organising themselves to form collaborative arrangements. A fish market owner says people need to join together: '*You are no good as single people. You can't fight anything on your own*' (KI-78). There are two main ways of organising: the first is to align with a strong national lobbying power such as the NFFO (KI-12). One fisher says this body has supported them fighting to transform their status quo:

> I've only been a member of the NFFO for a couple of years... and they've been excellent with us especially with this herring and drift net ban they represented us very well...we're lucky with the NFFO. I went to Brussels and the NFFO were there, Barrie Deas did a wonderful talk full of information, he'd done his research and it's thanks to people like that that we actually have a voice (KI-11).

Other communities affiliate themselves with the New Under Ten Fishermen's Association (NUTFA): '*At the moment most of the U-10m boats belong to NUTFA*' (KI-20).

The second way fishers organize themselves is locally, by forming fishing associations (FAs). A fisher refers to his FA as '*a voice which can stand up for you. To view your concerns... you need local guys*' (KI-98). An IFCA CFO says FAs enable communities to mobilise against common threats: '*you have an association because somebody's going to do something that you don't like...people will quite naturally come together against something...rather than coalesce for something*' (KI-108). A FLAG coordinator says FAs protect fishers from other marine users (KI-82). Another way FAs push for change is by representing their community in negotiations. For example, a fisher says his FA has '*been able to negotiate on behalf of the whole industry and secure payments for disruption*' (KI-83). A FA lead says FAs serve as conduits for fishers' protests to the government: '*if we want to remonstrate about some idea that the government has got, we will use the association*' (KI-1). One fisher claims his FA effectively has the power to reject proposed by-laws:

> We can as an association...say that is not going to pass with us. We are not going to be happy with that and ultimately it is very hard for them to enforce so if they don't have our backing, it is not going to be feasible... We had a meeting just before Christmas and there were a few by-laws proposed...but we rejected them as we didn't think they were necessary...so we chucked the by-laws out (KI-35).

An IFCA CFO describes a very active FA in his district: '*it's thriving. It's got lots of members. They meet* [formally] *twice a year, but really they meet every day, on the radio...that's a really strong organisation nowadays. They are all part of it and two of their members are full voting members of our IFCA*' (KI-107). A fisher says that in his community '*Most of the fishermen are part of the association*' (KI-64). A FA lead says '*When we arrange a meeting here through the fishermen's association there are always loads of fishermen*' (KI-26). The success and effectiveness of a FA is attributed to its leader's characteristics. A FA lead has to be trusted: '*You have to have a certain amount of trust that the person is sitting there is actually going to represent your interest*' (KI-16). Two fishers say '*You need a strong leader for this*' (KI-67) who '*knows the lingo to use*' (KI-37). According to a FA lead, an assertive nature is crucial: '*I am definitely heard, I definitely make myself heard*' (KI-100). A FA lead says '*if I don't like something I tell them about it*' (KI-82). Another FA lead says he has specialist negotiation skills: '*my full-time occupation was a full-time trade union official, looking after the dock workers...I'm a negotiator, I'm an arguer*' (KI-4). A leader must also be prepared to engage with external organisations (KI-98). Two fishers say FAs are more effective when they collaborate with each other: '*We're affiliated with the XX fishermen's association...the YY fishermen's association* [and]*...the ZZ fishermen's association, so we all talk to one another*' (KI-64); '*There is another fishing cooperative in AA and we do try to all work together. We share a vehicle that takes the fish down to BB and discuss our options together and work together to keep the costs down*' (KI-6). Another fisher says FAs also liaise with NUTFA: '*they are another voice that can shout for us if we need it*' (KI-65), while a FLAG coordinator says his FA is linked to NFFO: '*We have a relationship with the NFFO...and I am a director of the NFFO*' (KI-82). A FA lead says his association liaises with the Harbour Commissioners (KI-4). Another FA lead says his FA networks with local environmental groups on MCZs: '*We have the fishermen and the Wildlife Trust and all the stakeholders all working together. So we were one of the first lots to have our zones accepted because we worked together to achieve that*' (KI-38). A FA lead reports that his community has recognised the importance of effective FA leadership so much that it has given the leader a full-time professional post and turned the FA into a limited company (KI-81).

A FLAG coordinator says collaborative working relies on an ability to unite: '*The big strength we've got here is the unity*' (KI-81). According to a IFCA CFO, this unity is based on mutual support and trust: '*The fishing community...they don't agree with everything all the time, but in general terms they do. They stick together and there is trust*' (KI-106). Honesty between fishers is a key factor: '*I mean if we ever caught a pot by mistake I'll always haul it in and give it back*' (KI-18). Another key factor is having a common foe: a IFCA CFO says nothing unites a local community behind its FA more than an unpopular management decision: '*they can become very active if you try and do something in their area that they don't like*' (KI-109). A fisher agrees: '*when their backs are up against the wall, they do rally round and work together*' (KI-80). For example, a FA lead reports that the threat of a wind-farm on prime fishing grounds united his community FA so much that they came together in successful opposition:

> This was around a survey that a wind-farm wanted…to run right through their grounds… and they wanted to close this huge area in the summer months…There was a panic about this as they were being asked to shift their gear away from a really profitable area at the time of year that they make the money that sees them through the winter…there was a really arrogant approach from the developer, who seemed to think that they could just demand the sea cleared for them…People banded together and came up with a very robust response to that and ultimately it didn't happen (KI-81).

Another major success of a FA was to bypass fuel suppliers by creating a community-based fuel company:

> The fuel company was set up by the fishermen. It is owned by the fishermen. You can only be a shareholder in the fuel company while you're a member of the XX Fishermen's Association…the fishermen…receive every year…the profits, the dividends...the glue that holds our association together is our…fuel company…about 14 years ago, they [developers] were running a telephone cable…across to Belgium…and they wanted us to not fish an area…So we said yes, our boats will keep clear of that but it will cost you…we wanted a block sum, ten grand…and they said too much, too much…they said they would get it in within a week. But…we know the ground and it is hard chalk and flint...the flint is a killer and so we said no you won't. And they said yes we will…it was about £1,700 a week we negotiated. Well, a year later, they were still trying to put this telegraph line in. So needless to say we came out of it alright. So what we said to our chaps was that you can't have all the money…because we were desperate for a fuel company for fuel…we were having to buy the fuel from the barge down the harbour…but it was expensive buying the diesel that way. So we said to the lads, we'll set up our own fuel company and we'll have a tank on the quayside…So we set up the fuel company, successful after the first year or two (KI-64).

An IFCA CFO praises this innovative effort of the FA: '*they are bloody clever and they have used the model of all the fishermen owning the company that provides the fuel in the harbour and it is that company which provides the backbone behind their association, which is then the glue*' (KI-108).

A FA lead describes another radical effort of an FA to set up a not-for-profit consultancy organisation to negotiate their own compensation deals with wind farm companies rather than be dependent on private sector consultancy companies:

> There are consultants all around the country that will negotiate disruption payments for fishermen but they are usually then taking 10% of the payment for themselves. We don't do that, we add a fee on top of what the fishermen receive and bill the developer for that...We make life a lot easier for the developer: by dealing with us, they are speaking to just one person, instead of 70. Their lawyers don't have to get round 70 boats to get 70 agreements signed. It saves them a fortune just by dealing with us directly. So we bill them, rather than the fishermen. So the fishermen get everything as they are our members, so we are acting on behalf of them...we are a not-for-profit company...We simply need to pay our costs (KI-81).

An IFCA CFO describes another transformative strategy—a fisher training school set up by a FA with FLAG support to help recruitment into the local fleet (KI-109). The FA lead involved reports how this has been a great success:

> I was involved in the FLAG from the inception...And this area was awarded £1.35 million. With this we started the cooperative up, bought the land and the office on it, so we can now hold all the training on it for fishermen...We have issued something like 300 mandatory certificates for people who are just coming into the industry and didn't have any certificates...we have just finished two courses here last week and we are just about to start on two more courses, as there is a continuing demand (KI-4).

Radical transformative resilience also needs awkward or 'bolshie' fishers. A FA lead says: '*I definitely make myself heard...we have had some quite heated debates*' (KI-100). Another FA lead explains how he fought for a three-mile limit for MCZs: '*an MCZ...should be for our benefit...for the inshore grounds...I have been pushing for the three-mile limit...I will be a pain in their arse till I get it...a thorn in their sides...I put forward that we need a three-mile limit that you could only get access to if you were a beach-launched vessel. But my name was mud. I was called everything under the sun. But I couldn't care less, and I still want that to happen*' (KI-65). Another FA lead describes how he persisted in his opposition to MMO marine spatial planning decisions which threatened fishers' livelihoods:

> [In] the marine spatial planning process...they are arguing about which bit of the sea should we give to gravel extractors and which to energy companies...I argued quite strongly that fishing should be included in that too...we were there first. You are talking about which bit of the sea we are currently using you are going to give to someone else, effectively in perpetuity...our argument is that they should protect our interest as well...[Currently] the right to fish is trump-able by everything. So I would like something more secure in there. And I am fighting for that from the MMO (KI-82).

A fisher tells a hilarious story of how he used the press to play political parties off against each other in his campaign against the siting of a wind farm: '*you need to be able to play politics...I am a devious bastard at times. You got to be able to stand out*' (KI-73).

On managerial attitudes, some respondents call for a radical change in the attitude of managers at all levels towards the under-10m sector.

> If you want to go back to a working relationship with the industry then they need to look at… areas… like the south west and up the east coast where…fishery officers...have a totally different attitude. It isn't "I'm God and you will obey me, because I've got the right to do anything I like". There's an attitude of working with the industry and working with the fishermen cause there's a mutual concern…we're not all criminals as we're deemed to be in the south east of England…Until that mind-set is changed we are not going to get out of this…there needs to be a change of attitude within the hierarchy of the MMO…We need a briefing with the MMO officers about a code of conduct which should be laid out (KI-48).

A fisher says MMO managers have to understand whom they are managing: '*they are in their title managers, so if you are going to manage, you have to have individuals at the top who understand what they are managing*' (KI-21). Another fisher says managers should '*Work with the fishermen…Be more involved*' (KI-7). A fisher says good management entails empathising with the managed:

> it would be nice to have people in the MMO who have an understanding of the industry, who can talk to you and whom you can relate to...It would be a nice thing to think that you could ring someone up and have a talk with them about the fishing industry…and they would be like "Oh yer, I know"... [Currently] You ring them up and it's all "duh, duh, duh (KI-10).

Moderate transformative resilience is the strategy adopted by inshore fishers who seek to modify rather than overturn or replace the system. There are five elements in the moderate transformative strategy: rebalancing governance; reducing bureaucracy; improving communications; increasing quota flexibility; and using fishers' knowledge. On rebalancing governance, one element of this strategy is to adjust the governance system in order to give the inshore fleet a fairer share of decision-making. A fish merchant says '*Government need to be experimenting with ways of meeting with fishermen in a way that facilitates useful dialogue not just pushing the same format*' (KI-51). A fisher argues for equal representation on IFCA committees: '*If there are going to be two anglers, then you need two fishermen*' (KI-39). A fisher says there should be fishers from every port (KI-10): '*People have to reflect the industry*' (KI-4). A FA lead says they need candidates with local knowledge (KI-1). A fisher says people who should be on the committee are those who will be affected by its decisions: '*Have a look at who sits on the committee. There are people from all walks of life who whatever happens to the lobsters it doesn't mean anything to them. It won't change one penny in their pocket. Get the people who are going to be affected badly by the decisions and the fact that they may not be able to make a living any more*' (KI-21). On the other hand, an IFCA CFO urges committee members to strive towards the common good: '*you need people that can work toward finding common objectives and common goals and can work in a group and can understand the process of democracy*' (KI-113). A fish market owner suggests the inshore sector should pay fishers to attend committee meetings: '*it is very expensive for active fishermen to be representatives as it means losing a day at sea…the fishing industry itself should impose a levy, so that whoever is going to be the representative could be funded to attend these meetings. The industry would then expect fishermen to be able to represent them...that is a problem where the industry could do a lot more to help itself*' (KI-20). One community reported how they no longer waited for the IFCA to call meetings: '*they* [the under-10m fishers] *will tell our chairman and he will call a meeting and give us the opportunity to have our say*' (KI-36). They have also invited their IFCA CFO to be a formal part of their association meetings thus ensuring they are kept abreast of the IFCA's work and interests: '*I'm the secretary or the scribe as I call myself. I do all the minutes and note-taking and everything*' (KI-106).

A FA lead recommends that the IFCA regains autonomy over its selection process by removing the '*power to choose IFCA committee appointees away from the MMO back into the hands of the IFCA bodies...I don't understand why the MMO has the whip hand*' (KI-48). Another FA lead rejects MMO's idea that the role of members of committees is to provide their expertise, not to represent their interests: '*Fishermen need representation in their own right and their own voice*' (KI-2). A fish merchant proposes abolishing voting in IFCA committees because fishers are always outnumbered: '*Voting shouldn't be there at all...if you have to vote on something...the industry will always lose as it is a numbers' game. How can four people outvote the other 18? The other 18 always go along with what the expert says* [yet]...*the only thing he is an expert in is manipulation*' (KI-51). Good decisions are not reached by majority vote but by discussion that is sufficiently prolonged to reach consensus.

A FA lead says a further way to re-balance the governance system for their community is for consultations to be collective exercises rather than individual one-to-ones, because fishers feel intimidated by lone interactions with regulators:

> When they consulted on the most recent by-laws, they wouldn't hold a public meeting: they held individual meetings with individual skippers and that is all that they would have...You have two guys in uniform, one of whom is in charge of enforcement and you as a skipper in an office. And they are asking you if you agree with the new laws they want to bring in. Now that is not going to get a...frank debate is it? "Oh, aren't you the chap who can inspect my catch and stop me going to sea? Maybe I won't disagree with you, then"...What they should have done is have a meeting in the town hall (KI-81).

However, a fish merchant says fishers find public meetings intimidating: '*Those people still have some genius ideas and they don't feel that they are eloquent enough to put those ideas forward and they get very self-conscious in these meetings...we should not have group meetings as the fishermen absolutely tie themselves in knots*' (KI-51).

Some respondents say they have seen real changes in policies as a result of their input. For example, one fisher says: '*The XX fishermen said that the information that the IFCA were using was from the 1970s, so was no good. And we asked why are we amber-based on that, so they changed it to a green*' (KI-102). An IFCA CFO claims that fishers' involvement in IFCA decision-making about an MCZ was genuinely a two-way process:

> A classic example is the way in which we set up the consultation on marine conservation zones...we set up a meeting and did a presentation and discussed what...Defra were expecting of us and invited all the fishermen...offering them a pasty and a pint if they came along.

> So they all turned up and we explained to them that we didn't have a choice: Defra imposed this on us, we've got to do something about the marine environment and protecting it, this is what we're suggesting and they bought it. They loved it. They said, "Ah that's brilliant". And they came up with their own ideas about voluntary codes of practice, which again, we've incorporated into the management structure. So it's been a two-way process: they have been fully supportive all the way…Defra [said]…several times in meetings that I've been to in London that we are the only IFCA in the country that had total support from environmentalists, water users, fishermen, and the local community (KI-107).

A fisher says '*You can approach them and put your thoughts forward and they will take what you say on board*' (KI-91). Another fisher says their views are taken more seriously than in the past because Defra now audits IFCAs' proceedings:

> I have just discovered that objections to things are really taken seriously. In previous times, when you wrote objections they would just go, "Oh yer, how sad". Now they need to tell Defra what you said, how they answered you and if you were satisfied at the end of the process, and if not, why not? I think this is brilliant, as even if you get beaten down at the end, you know that they have had to consider it and they need to come up with a good reason why they didn't consider your objection...It's a vast improvement (KI-39).

An IFCA CFO says that on some issues, fishers' views could be decisive: '*we have a good record, not of being at their beck and call and accepting what they say willy-nilly, but of listening to them and…implementing management that is appropriate in the light of the information we've been given*' (KI-111). Another IFCA CFO explains how he withdrew a proposed by-law because it was opposed by fishers: '*I tried to introduce a by-law, which would tighten things up a little bit and they rejected it. Oh no, we don't need it. Everything's working perfectly, why bring a by-law in when it's not necessary. So I said well, fine*' (KI-107). A FLAG coordinator says all they ask their IFCA is to '*Show us that they are listening to us. They don't have to implement everything, but…they should initiate measures that we have suggested to them, to help protect our industry*' (KI-66). An FA lead says:

> All you want is a fair crack. You just want to be listened to and seen to be listened to. Not just lip service. We had a whole list of suggestions that we gave them. You don't have to implement them all. Just take some…and start the ball rolling so that you show folks that you are listening. If you start to gain the trust of people, then you are moving in the right direction again (KI-48).

On reducing bureaucracy, a fisher says the management system should be made more intelligible to fishers by adopting layman's terminology: '*It would be better if they explained it more simply...just don't complicate issues but put it straightforward like...You have to go through so much nonsense to get there it gets on your nerves when things could be so much simpler*' (KI-37). An IFCA CFO agrees:

> You're talking about some quite complicated ecological concepts and they don't understand how that comes into their business or really affects them...And you're just left with this wealth of acronyms...and things that people have come up with to assess things which are pretty fictitious really…there is a lot that people expect them to understand as small businesses and if they don't, they feel as though they are completely disenfranchised. So a lot of it is about finding the language and the skills (KI-108).

Another IFCA CFO acknowledges that simplicity is vital to enable fishers to know what their obligations are: '*Not because fishermen are simple folk, far from it, but because it's a complex thing and it needs simple rules in order for people to understand that they are fishing in step with* [regulations] *and exploiting in a sustainable manner... it's the old keep-it-simple, stupid.*' (KI-105). Respondents also want managers to facilitate long-term planning. A FA lead says '*Even if they gave you a week at least you could try and organise your life around that*' (KI-12). A IFCA CFO agrees: '*if I was a fisherman, I would want to go "OK, well what's my quota going to be in two years' time? What is it going to be in three years' time? Well, you give me that model and I can start to develop my business model on those predictions*"' (KI-108).

On improving communication, many of the above bureaucratic flaws are symptomatic of a wider failure that the respondents attribute to the IFCAs and the MMO in not communicating effectively with fishers. On the IFCAs, a fisher says '*They* [the IFCA] *should communicate more, not just with their representatives. Sometimes we don't hear from them in months*' (KI-74). A FA lead says it is essential that communication caters to the particular needs of the fishers: '*We're a section of society that struggles to engage ... in normal methods or ... through normal mediums ... You have a responsibility to find the best way to engage with them ... For example, there are issues to do with adult literacy. What you don't do is end up with a 40-page consultation document that people you are trying to consult with can't read ... I think that there are fundamental flaws in the way the government engages*' (KI-15). Two fishers urge the MMO to meet them face-to face instead of relying on electronic communication (KI-96) (KI-111). An FA lead urges IFCA officers to get out amongst the fishers and take the flack:

> When it comes to consulting fishermen on new management measures, they should be less afraid of talking to them en masse...Yes, they would have people disagreeing with them, but that is their job. When we have meetings with all our members, I accept that it is my job to stand up in front of everyone and get shouted at. That's life. When you are in charge of an organisation you need to do that. As an IFCA they are making decisions that are going to impact on people's livelihoods, so they need to stand up and get yelled at. And they will find that... there will be a lot less shouting than they think there will be. People will get heated as they are passionate, but most won't, and quite a few people will agree with them and it will allow for a debate (KI 81).

However, other participants say they have no concerns about communication with their local management. One fisher says his IFCA CFO communicates with the movers and shakers in each port in the IFCA district: '*In every port, there are two or three people that run it and care.* [The CFO] *has made a point of going to every port and finding that guy or guys in every port, who will tell him what is actually going on*' (KI-39). Another IFCA CFO claims he holds 12 meetings a year: '*I was quite aware that if I'm not careful my position gets quite isolated from the reality and people don't talk to me and I don't see people, and I'm going on gut feeling. I don't like that. I'm not a control freak, but I like to understand what is going on out there...I hold twelve meetings around the coast*' (KI-104). He explains how he goes to great lengths to ensure that disadvantaged fishers can access information:

> One of the big issues...is that a lot of them are educationally bad...and a lot of them seem to suffer with dyslexia....So unofficially when they bring their statistics here, the girls will

> go through and make sure they have filled it in properly. Now we shouldn't really do that, but we're a public service, paid for by public money...All our documentation and reports are in Century Gothic type…because if you're dyslexic, it's one of the easiest types to be able to read (KI-104).

A fellow IFCA officer is equally sensitive to the needs of quiet fishers:

> Some people shout loudly and aren't affected. Some people are very quiet and are really affected. We try to get an information exchange which is not based simply on who shouts the most...One of the ways I really like…is evidence that is not emotional. It's factual. So let's talk about how you use this site. Let's go down to your boat and look at the plotter…. if you're the loud guy in the corner shouting and there's the quiet guy in the corner, thinking about how he is going out of business, then it's the quiet guy that we are going to go and deal with (KI-112).

A fisher says his IFCA is very approachable (KI-35). Another fisher says his IFCA's CFO '*has made a huge difference and has made a big effort to go out and meet people who were disenfranchised*' (KI-40). An IFCA CFO claims fishers already know of the many channels of communication available to them: '*They all know how to contact us. We are available by email, phone, walk in off the street, websites, through our members*' (KI-104). Another IFCA CFO reports that as a result of improved communication he receives far fewer objections to his proposals:

> A lot more consultation on the management measures and lots of events in ports along the coast, asking for peoples' views and what people would like us to prioritise [pays off]. I know from the new management in XX, that it met with a big positive response and we have advertised by-laws and the number of objections has been very low…that is because [in] these pre-consultations [we] met with people and listened to their issues and then amended the rules in line with these (KI-40).

A fisher praises his IFCA's use of drop-in sessions as a way of communicating in their area: '*This IFCA is really good at communicating and also pulling people in. They have had a few of these drop-ins ... it acts like a social event and it pulls them* [fishers] *all together...if only* [because] *they are scared that they might be missing something*' (KI-39). Likewise, a FA lead says his communication with the MMO was fruitful: '*I took the view that you get nowhere unless you're talking with people. I wrote to the MMO…last year, saying I didn't like their decision to discontinue the skate fishing…and they said can we come along and see you and so we had a meeting*' (KI-4).

On quota flexibility, respondents who adopt a moderate transformational resilience strategy propose modifying (rather than replacing) the quota system to make it more flexible. A fish merchant says: '*please remember that the inshore fleet, has to have flexibility. If you don't you might as well lock them up and throw away the key. 'Cause if you don't have flexibility, you can't react to the seasons*' (KI-51). For example, one fisher supports lengthening the short-term monthly allocations to longer time periods: '*If they gave us a year's supply of quota, then we could decide when the market is right, when the weather is right.* [This would] *give the fisherman a bit of flexibility to really plan ahead and manage their own fish*' (KI-58). Another fisher praises the increased flexibility shown when the MMO temporarily allowed fishers access to their three months' quota at once (KI-31). A FA lead urges management to give fishers advance warning of upcoming changes to regulations to enable them to plan their businesses efficiently: '*Even if they gave you a week at least you could try and organise your life around that*' (KI-12).

Another fisher says the quota distribution needs to be decentralised to allow allocations to be set in line with local stocks: '*the quotas need to be realigned with the availability of fish, rather than the availability of paper fish. Need to be based on fact. The quota is out of step with reality*' (KI-21). A fish merchant explains that a decentralised system would allow fishers to exploit sudden surges in stocks: '*the opportunities from healthy stocks are negated if you don't have access to them through access to quota. So you could have soles knee deep, but if the quotas stay as they are, you still can't touch them. And you can't discard them, so what are you going to do with them?...this is where your flexibility comes into it and this is what I am advocating*' (KI-51).

Another moderate transformative change demanded by respondents is for fishers' knowledge to be incorporated into the science which underlies their regulations: '*The fishing industry has to be more involved with the science*' (KI-4). One such example given relates to stock assessments: according to an IFCA CFO, the existing system of stock assessment is deeply flawed: '*the MMO and Defra should look more carefully at the scientific analysis ... It's all very well saying there's evidence that the fish stocks are becoming depleted but the fishermen will challenge that...there's loads of cod...out in the south-west approaches, but there's a limit on the amount they can catch. So I think the biggest issue is weak interpretation of scientific data largely from the MMO*' (KI-107). A fisher claims management should use fishers' knowledge: '*The fishermen know more than the scientists about how many fish are about and about how some of the stocks are increasing quite well*' (KI-26). Another fisher says

> there is knowledge within the fishing industry. This could be utilised in a really effective way...We can be working with them [scientists], be giving them all the knowledge we have, we could be doing good scientific work...given that we are out there every day observing the stocks, we would be ideal to use to monitor fishing effort and fishing strategies and fishing policy. But instead we are treated as if we are numpties with no brains in our heads (KI-21).

Another fisher reports vainly asking to join scientists doing fisheries research: '*we have asked them on several occasions, that every time there is a* [scientific] *survey if we can have a representative out on the boat. And they have said "Yes that's fine", but we still see the boat out there surveying, and they don't ask us*' (KI-59).

However, a market owner claims collaboration between fishers and scientists is already taking place: '*There have been very encouraging partnerships between the fishermen and the scientists and that has to be the way forward*' (KI-20). A FA lead says such collaboration is happening in the cockle fishery (KI-4), and a fisher describes a collaborative project he was involved in researching into spur-dogs (KI-71).

An IFCA CFO testifies to the value of such collaboration in a project on rocky reefs:

> We brought in a by-law prohibiting mobile gear on rocky reefs...It was ultra-cautious and very precautionary, and therefore in closing down complete areas...to bottom towed gear, there are areas which we have therefore closed down which aren't rocky reef, in other words, sand and muddy ground etc., to which that prohibition shouldn't apply. Therefore, we reached out to the local fleet and said "Look, we've made this by-law, it's what we felt we've got to do but if we can we will open up areas in the future which are not rocky...[but] are sandy…Can you help us with this?" And the industry were able to come forward very effectively and point us in the direction where we would commence our ground-truthing and other research activities to try and establish those areas, with a view in due course, if we can, to open up areas to trawling activity to resume there. The knowledge that was displayed to me about their grounds was tremendous (KI-111).

Another IFCA officer testifies to the high quality of data obtained by a collaborative project on MCZs: '*We were the only IFCA in the country that got five out of five, the top marks for every single sub-site in our marine conservation zone. So we got a gold star for that…a lot of it was* [because] *the fishermen helped us. They're brilliant*' (KI-107). A FA lead says fishers carry out research that scientists cannot:

> We collaborate with them on our research program. We have been clear with them from the start that our research program is only ever any good if the data get used. It needs to be used by the IFCA as they are the ones that have the power to manage things. So we share our data with them. We are engaged in a number of collaborative projects with them. We had a meeting with their scientific officer yesterday about a range of projects. There are things we can do that they can't. They don't have a boat down this end of their remit…we have a research boat here and if we want the IFCA to manage this area well, then we'd better help them out getting some data. We are not-for-profit so we can do things quicker and cheaper than they can. We can be reactive. They haven't hired us: we work in partnership instead. We have worked on some joint grant applications together, but they don't hire the boat from us. A lot of the stuff they say would be useful to them we can easily collect whilst we are doing something else. The stock assessment stuff, that's the mainstay of what we do and we just give them those data. The IFCA have measured 7,000 lobsters last year as part of their boarding duties. This last year we measured 20,000…in an organised sampling program with regular sampling projects, with very good comparability of data. We just have the capacity to do that and they don't have it. They are facing big budget cuts and they have a huge problem with staff retention. We can just get stuff done…we were asked by the IFCA yesterday to conduct a study into how long it takes them [lobsters] to regrow their limbs after they shed them…Great idea and it came straight from the IFCA and it's a great example of working with them (KI-12).

Another FA lead claims fishers' research data are more reliable than those of Cefas: '*We have the fishery industry group which is unique to the country and they have their own research vessel which does research on behalf of the fishermen. They have got more robust data than Cefas as to what is going on up here. We found that out at the shellfish conference down in Greenwich...the more that Cefas talked, the more we realised that the data that Mike had round here is more robust than what they got*' (KI-81). A FLAG coordinator says they are undertaking their own research to prove how sustainable their fishing is: '*We are doing our own research to check if it* [our fishing] *is or isn't* [sustainable] *and if it isn't then we will do something about it. But the fact that there is a willingness to find out if it is sustainable or not, is in itself a strength*' (KI-82). A FA lead says: '*We've got records and anecdotal things going back donkeys' years... we do observe what's going on around us*' (KI-3).

Collaboration not only takes place on ecological data but also on economic data, as another IFCA officer reports:

> We've got a monthly survey which fishermen run but we process [their data] that looks at what they catch, where they catch it and how many people [are involved] and what the price of fish are...We have been running that for two years. The fishermen do it all and it's all anonymous but what we do is we provide the infrastructure, we pay for the fishermen's liaison person and we then process all the stuff and then give it back to them...So we've got two years of data there and that's really good...[It] gives a really clear economic model and what it shows is that the good inshore fishermen are working on optimal values in a fishery (KI-108).

Scholtens et al. (2019, p. 326) may be interpreted as summing up the difference between a moderate transformative strategy and a radical transformative strategy as the difference between a 'harmonious' approach and an 'antagonistic' approach:

> In terms of empowering small-scale fishers for gaining resource access, the difference between these two approaches is significant. A harmonious approach to empowerment may entail building institutional capacity among fisher organizations and facilitating the negotiation of collective outcomes with competing resource users. An antagonistic approach may rather entail confronting exclusion, inviting public protests and challenging rights and discourses that privilege dominant resource users.

5.5 Conclusion

This chapter on respondents' perceptions of the way English SSF have dealt with the threats and pressures facing them has shown that three resilience strategies can be identified: passive resilience; adaptive resilience; and transformative resilience. Within each of these three strategies, we have found distinctions: passive resilients may be pessimistic or optimistic; adaptive resilients may make strategic or tactical choices; and transformative resilients may adopt radical or moderate approaches. There are, in fact, a myriad of different ways in which fishers may respond to the stresses of the situations that confront them. What is striking is the predominance of transformative strategies over passive and adaptive strategies. This does not necessarily mean that more small-scale fishers (SSFs) are engaged in transformative strategies than in passive or adaptive strategies. It may mean that transformative strategies are more vivid in the minds of SSFs, and that when they pay off, such strategies exert a more significant impact on the lives of SSFs. It could also reflect SSFs' greater desire to change their seascapes than to accept or adapt to them. Whatever the explanation, it is encouraging to see SSFs' enthusiasm for transformative initiatives, because it suggests that the English SSF has the potential not only to survive, but to flourish, especially if it is accorded fair treatment by agencies tasked with its regulation.

Chapter 6
Discussion of the Resilience Strategies

'Small-scale fishers are becoming increasingly marginalised' (Maarten Bavinck)
'What is needed is the large-scale promotion of the small-scale' (Patricia Pitchon)
'I'm not saying big is always bad, but when it is bad, it is very, very bad' (Anonymous)

6.1 Introduction

The previous chapter has demonstrated how English inshore fishers have coped with the many challenges they have faced during the last 30 years. Some fishers have dropped out of the sector, but the remainder have adopted a variety of resilience strategies in order to survive, including passive, adaptive and transformative responses. In this discussion chapter, we seek to understand the drivers and implications of these resilience strategies by addressing five key questions: What are the factors that determine English SSF's choice of resilience strategies? What is the role of Inshore Fisheries Conservation Authorities (IFCAs) in influencing the choice of those resilience strategies? Can different resilience strategies co-exist? Are there any trends in resilience strategies? Are some resilience strategies more sustainable than others?

6.2 What Are the Factors that Determine English SSF's Choice of Resilience Strategies?

Fishers' choices of resilience strategies depend partly on factors within themselves (psychological factors) and partly on factors outside themselves (e.g., environmental, social and government factors). We discuss the drivers of these factors for each

© The Editor(s) (if applicable) and The Author(s), under exclusive license to Springer Nature Switzerland AG 2021

R. Korda et al., *Resilience in the English Small-Scale Fishery*,
https://doi.org/10.1007/978-3-030-54245-0_6

of the three strategies in turn. The passive resilience strategy in part can be attributed to distinctive personality traits. Some inshore fishers exhibit an embedded or innate self-reliance, stubbornness, pride and independence, manifested in lone behaviour and unwillingness to try and exploit change. Their stance is a form of defiance or determination that they would keep going, as they always have, in the face of adversity and threats to their livelihoods. Paladines (2015) sees this mentality in Ecuadorian fishers. In the English case, this means some fishers denouncing and refusing on principle to take advantage of the quota leasing system, even if others deem it a lifeline. In some communities, older generations of fishers see the younger generations leaving the sector to move to more stable, profitable and less stressful industries such as offshore renewables, but they themselves refuse to jump ship not because of an inability to diversify but because of a stoic refusal to leave the industry that they love. They see their job as a vocation, a way of life and ultimately a validation of how they define themselves, and in their eyes, there is never any question of doing anything differently even if staying in fishing means suffering increasing economic hardship. Ross (2013, p. 57) says "The importance of this identity may help to explain why fishers continue to fish, even when it is no longer economically viable to do so". Other passive resilients are characterised by Pollnac and Poggie (2008, p. 199) as 'addicts': "A commercial crabber from Alaska said, 'As any fisherman's wife will tell you, fishing is an addiction. And for commercial fishermen, consider it a gambling addiction'. This is an insightful observation, fishing *is* like an addiction, and most fishermen would do anything to avoid the potentially painful withdrawal symptoms" [italics in original]. Others feel a duty to protect their local marine heritage, to honour their heroic predecessors, and to repay the sacrifices made by those who came before them by continuing with these practices, thereby maintaining a cottage industry in the face of a global market, even when it becomes personally uneconomical. Such participants say they do not wish to change their methods as they are committed to upholding traditional practices even if these have become financially untenable, not least because these practices are environmentally friendly with extremely light marine footprints. Some are influenced by less positive factors, remaining in the industry because they lacked alternative options. External factors leading SSFs to choose passive resilience include the negative response of the authorities to the plight of struggling fishers. If the authorities respond unhelpfully, giving little or no ground to individual fishers or their communities who are on the edge of economic collapse, adaptive and transformative strategies may appear futile, and the only feasible strategy is to become passive and disengaged (Henry and Johnson 2015).

The main psychological characteristic of individuals and communities who choose an adaptive resilience strategy is flexibility and the willingness to experiment. Adaptive resilients repeatedly adjust their working practices in response to changing external drivers, thus enabling them to continue operating, maintaining their identity, function, and structure in response to feedback (Pearson and Pearson 2012; Wilson et al. 2013). This is a continuous, incremental process allowing fishers and their communities to live with ever-changing circumstances and uncertainty and is said to be embedded in the inshore fleets' genes. It is a strategy similar to that described by Allison and Ellis (2001, p. 384) as the sustainable livelihoods approach, which involves the use of low-tech fishing gear to keep down costs and maintain flexibility: "A low technology, opportunistic, fishery adjusts its behaviour to the available resource: if the resource declines then the fishing effort is scaled back and family resources turn to doing other things". It is reliant on the existence of an environmental, legal and economic landscape which facilitates flexibility and experimentation (Anderies et al. 2006). The type of person who adopts such tactics is one whom embraces rather than fears change; is positive rather than negative; is constantly on the lookout for new challenges; does not wallow in the past but looks to the future; is decisive in abandoning past practices that are redundant; and is perpetually innovating and willing to take a chance on failing (Gunderson and Holling 2002; Armitage 2005). Pollnac and Poggie (2008, p. 194)'s psychological profile of fishers seems to fit the adaptive resilient personality: "the relatively risky nature of the occupation of fishing attracts and holds individuals manifesting an active, adventurous, aggressive, and courageous personality". They are also able to learn from their own and their fellow fishers' failures when they are experimenting, and they possess the ability to plan ahead (Folke et al. 2003).

However, there are limits to the magnitude of shocks and changes that an ecosystem or community or individual can absorb and still maintain its adaptive capacities (Béné 2013). Béné and Friend (2011, 130) argue that "the *exposure* and *sensitivity* of fisherfolks to risks are relatively high (in comparison to other socio-economic groups) and their *adaptive capacity* is generally low". Coulthard and Britton (2015, pp. 285, 286) point out that some fishers may have little room for adaptive strategies, or may have to pay a heavy price for adopting them: "people have a limited degree of agency in negotiating their own adaptation strategies…a breakdown in family relations, as a consequence of adaptive strategies, point to 'mal-adaptations' that could lead to a collapse of both the adaptation strategy and the livelihood itself". Moreover, Coulthard (2012, p. 3) says adaptation may be a mixed blessing: "adaptation strategies, although reducing risk on the one hand, may also incur forms of harm and ill-being that need to be accounted for in policies that promote adaptation as a means of risk reduction". Furthermore, pressures may occur that are so large they overwhelm the adaptive capacity of the fishers or their communities, and result in one of two responses—passivity or transformation.

Transformative resilients hold to a belief that it is possible not only to adapt to change, but also to bring about change (Walker et al. 2006; Folke et al. 2010)—i.e., to reshape the world in a permanent way and thereby regain control of their lives. Armitage et al. (2017b, p. 262) describes transformative resilience as a political exercise: "transformations of the coastal commons are inherently political, and shaped by contestations, [and] inequitable relations of power". By contrast to adaptive resilience, transformative resilience is more often a collective than an individual strategy, and while most transformative initiatives are local strategies, there are also a handful of national strategies in which fishers attach themselves to campaigns such as Brexit, designed to disentangle the UK from the EU's CFP and reclaim national power over fisheries. At a community level, factors encouraging fishers to opt for transformative resilience include the actions of key individuals embedded within the community who possess leadership qualities, and the ability of communities to self-organise and to maintain good links with outside networks. According to Schwarz et al. (2011), leadership at the local level has a key role to play in unlocking the capacities of communities to pursue transformative changes (Schwarz et al. 2011). Acts of transformative resilience in the English inshore sector are usually spearheaded by one or two key individuals within a community who are energised and able to achieve support: they work with such vigour that other fishers cling to and support their cause. People tend to group around certain ideas based on the beliefs of those whom they trust and respect, compounded in some cases by a shared history. Often, the prime leadership figure is the head of a community's fishing association. The capacity of such leaders to be innovative enough to imagine an alternative future—to think outside the box—and daring enough to act on that vision, is a crucial factor, as is the personality of the leaders and their ability to be articulate, honest, inclusive and transparent. When community leaders demonstrate these qualities, their communities draw together in the belief that an alternative future is possible. In this situation, individuals are prepared to work on a less individualistic basis, and a collective spirit of 'we are all in it together' emerges. These social cascades could facilitate broader social changes challenging the status quo. The ability of a community to self-organise is also critical to transformative resilience, because there are limits to an individual's capacity to respond single-handedly to a crisis, and collaboration and a pooling of resources is required to overcome these limits. But such collaboration relies on the ability of fishers to unite which requires mutual trust, respect and support: communities which are riven by hostility or distrust are unlikely to generate transformative actions.

Links with outside networks is vital to this process, given the complex mix of linkages across a multilevel scale required to support transformative schemes (Folke et al. 2003). Leaders must be able to develop and maintain supportive external networks at a local, regional and national level: according to commentaries, systems that have more open feedbacks—that is, those connected to external networks, such as local, national or international organisations—or external financial systems such as the FLAG network, have an increased chance of crossing transformative thresholds (Tompkins and Adger 2004; Anderies et al. 2006). In the English U-10 case, networking occurs in three main ways. The first way is to make use of a strong national lobbying power. For example, respondents speak of how they have networked with the National Federation of Fishermen's Organisations (NFFO) to overturn an EU drift net ban. Other communities have affiliated themselves with New Under Ten Fishermen's Association (NUTFA) to overturn the quota allocation crisis through a judicial review and an inshore producers' organisation (PO) campaign. Some participants try to transform the political landscape by networking with political parties such as UKIP (United Kingdom Independence Party) and the Brexit Party in the campaign to leave the EU. The second type of networking is with neighbouring communities to deal with common concerns. This is illustrated by the linking of communities to adopt the community quota program in an attempt to transcend the quota allocation crisis. Another example is the innovative way some communities have networked together to obtain Fisheries Local Action Group (FLAG) funding for joint or complementary projects. The third type of local networking is with IFCAs. For example, heads of fishing associations put themselves forward for membership of their IFCA committees, and some respondents explain how they see real changes to policies as a result of their input, such as amendments to a marine conservation zone (MCZ) proposal to allow a fleet to continue working within a contentious area, overturning the threat of their IFCA adopting a precautionary approach.

However, transformative strategies face formidable obstacles. For one thing, they are major exercises to mount: it takes more effort to transform a system than to adapt. Driving a transformative strategy forward is a resource-intensive process and very risky, as should it fail, it could create a negative feeling of apathy amongst the community. Also, transformative strategies demand a high degree of commitment from SSFs, which is not always forthcoming. For example, most respondents call for a radical transformation of the quota system, but only a few communities actively push for it, like Whitehaven and Dungeness who put themselves forward for the community quota scheme. Also, many respondents demand a redistribution of the quota from the offshore to the inshore sector, suggesting ideas such as re-nationalising quota to provide a fairer share for the SSF, but only a couple of communities address this by backing action through a judicial review. Again, many communities express interest in creating an inshore PO, but only Hastings and Ramsgate pursue a proposal to establish an inshore fisher-run PO to operate at national level to obtain the same benefits as the offshore industrial POs, and only a small minority of SSF have joined the Coastal PO. Also, while many fishers want more self-governance, this is an idea implemented by only a small pool of communities.

Finally, the success of a transformative strategy may depend on the way it is received by the system it sets out to transform (Lebel et al. 2006). Failure by management systems to recognise the importance of these transformative initiatives and respond to them in a respectful and positive fashion is a major obstacle to their coming to fruition. This is witnessed in a community in the south east district, where fishers claim they face a negative reaction from the Marine Management Organisation (MMO), though very positive reactions from other external groups such as the local council, non-governmental organisations (NGOs) and universities enables them to persist with their initiative. By contrast, in a community in the north east district, the fishers' drive to prove their fleet sustainable and set up their own research system, is welcomed by their IFCA.

This leads us to discuss in more detail the crucial role of IFCAs as a factor in determining SSF's choice of resilience strategies.

6.3 What Is the Role of IFCAs in Influencing the Choice of Resilience Strategies?

In 2011, when the IFCAs replaced the SFCs, they were handed an additional conservation mantle, which came with hefty workloads imposed by Defra together with tight timeframes. One area where some IFCAs saved resources was by moving to a more reactive or inactive engagement and communication mode, whereas others continued to prioritise active mechanisms despite the drain on their resources. These decisions were found to influence the choice of resilience strategies pursued by fishers and their communities. Symes (2014) claims that IFCAs are in a good position to encourage resilience in the SSF, and while many believe they do so, our study shows a mixed picture. We have placed the ten IFCAs on a continuum of facilitating the resilience strategies of the fishing communities within their jurisdiction based on their engagement arrangements at the time of research—from proactive to reactive to inactive. In order to maintain anonymity, the IFCAs were assigned codes and henceforth will be referred to as IFCAs One to Ten. The most proactive are IFCA Two, IFCA Three and IFCA Five, all of which pursue proactive engagement mechanisms, thereby helping to promote the adaptive, and (to a lesser extent) transformative, resilience initiatives of individuals and communities. IFCA Ten, IFCA Nine and IFCA Seven are found to pursue reactive arrangements, setting up meetings as and when required and responding to fishers' overtures but not initiating action. The IFCAs pursuing the most inactive arrangements are IFCA Six, IFCA Four, IFCA Eight and IFCA One, which have moved their communication strategies almost entirely online and are slow to make responses to fishers' resilience approaches.

One of the most proactive IFCAs is IFCA Two, where the Chief Fisheries Officer (CFO) has been a dynamic part of the community for over 30 years during which he has gained a deep understanding of district's fishing communities. He is careful to integrate the IFCA into the community to the point of attending fishing association's meetings; providing sea transport for fishers in remote areas to attend IFCA meetings at times convenient to them; and taking proposed management measures to the fleet before putting them to the IFCA committee. On many issues, including data gathering, assessing the need for a pot limitation, MPA management requirements, v-notching lobsters for stock assessment purposes, and banning the landing of berried lobster hens, this partnership arrangement based on mutual trust works well, building fishers' confidence in their ability to protect their own waters and keeping the community united. As a result, IFCA Two is able to meet its environmental objectives whilst at the same time promoting adaptive resilience strategies, and even supporting some transformative strategies, adopted by its fleet.

IFCA Three also pursues a proactive engagement strategy, working in an inclusive manner to ensure that fishers in its district play a genuine part in the decision-making process on issues affecting them. The IFCA tries to get a broad range of fisheries and geographic areas represented on its committee by approaching relevant communities and encouraging them to nominate possible candidates. Also, when major policy changes are mooted by Defra, IFCA Three holds meetings around the coast targeting all communities rather than a select few, in a variety of formats based on each community's needs (e.g.one-on-one meetings, open meetings, focus group meetings) and at locations and times convenient to the fleet, drawing people in with the promise of pints and pies. For example, the IFCA CFO invited fishers to a meeting to point out the gaps in its data on a biogenic reefed area requiring management, an exercise that allowed some fishing around (though not within) the biogenic reef area, thus satisfying both the Marine Conservation Society (MCS) who are pushing for a full-site approach, and the fleet. Fishers within the district report a positive relationship with their IFCA, giving examples of collaborative work and praising it for encouraging fishers to adopt adaptive resilience strategies. The CFO is keen to improve communications with poorly educated fishers, and with dyslexic and older fishers. He claims that the IFCA itself has an adaptive approach to fisheries management: *'Adaptive management is the other element that we are very keen on here for managing fisheries...We are creating a statutory management committee for adaptive management'* (KI-104).

IFCA Five's proactive work is also driven by its CFO, who organises meetings with fishers on a one-to-one basis and runs a series of drop-in sessions as social gatherings. This results in increased attendance and brings fishers into discussions which ensure they are briefed about the evidence on which management measures are based. IFCA Five's CFO has also set up an informal fishers' council which sits alongside the IFCA committee and acts as an advisory board, thus responding to the charge of exclusivity. This IFCA has commissioned fishers to collect evidence to confirm that the portion of a proposed closed area that is of most interest to fishers do not contain any features that need protection. The CFO responded to the fishers' plea for protection against gang-led illegal involvement with a clam fishery, turning the problem around almost overnight by the introduction of a permitting scheme. Fishers also appreciate the role of the IFCA in providing them with new livelihood opportunities in aquaculture. One respondent commends IFCA Five for being pragmatic in not adopting a purely environmentalist approach: *'On the whole, in our IFCA...the chief fisheries officer and his team are trying to hold back a flood of environmentalists who are putting enormous pressure on fishermen. Their mantra is that the fishermen are clambering all over the seabed and taking every living thing out of the sea. But our IFCA know that isn't true and they do want to get on with everyone. They have a lot of pressure on them from the environmentalists with the conservation zones, but they seem to take a pragmatic take on it'* (KI-41). Another respondent praises IFCA Five's CFO for his inclusive management: *'I am now on my 4th Chief Officer and this one is the best. He is brilliant...He gets feedback from people on the ground and he deals with it...this IFCA has helped as they are a point of contact that all the fishermen have to have'* (KI-39). IFCA Five's CFO describes the authority as a co-management system (*Fishing News* 14.11.19, p. 11). In Defra's report on IFCA Five, stakeholders' comments were equally positive: "Respondents commented that IFCA staff are pragmatic, professional and make good efforts to engage with stakeholders" (Defra 2015, p. 48). All of this indicates that IFCA Five actively helps fishers to adopt adaptive resilience strategies.

The second group of IFCAs, the reactivists, claim they are committed to proactive engagement with their inshore fleets, but say they are thwarted by the burden of work handed down by Defra. They have reduced their face-to-face engagement with fishers to holding a few general meetings when needed in key communities across the district and responding rather than initiating—an approach which risks favouring the better organised and more vocal communities. IFCA Ten's reactive stance was illustrated by their handling of Defra's shift in policy regarding management of fisheries in European protected sites: they held meetings with fishers likely to be affected in order to obtain the required evidence to justify opening back-up areas so trawling activity could resume there. These are encouragements for adaptive strategies. One respondent says fishers are able to communicate their concerns to IFCA Ten and feel that their views are taken into account: '*You can approach them and put your thoughts forward and they will take what you say on board*' (KI-91). However, despite this reactive engagement with fishers, some fishers criticise IFCA Ten for favouring some groups over others as they rely on fishers coming to them, and for stopping communicating by paper, thereby privileging those fishers who are comfortable with electronic communication. In Defra's review of IFCA Ten, stakeholders' comments were very positive: "Engagement with stakeholders was commended" (Defra 2015, p. 39).

IFCA Nine also reacts positively to its communities' overtures, keen to act on issues brought to their attention by local fishers, thereby encouraging them to adopt adaptive strategies. One such issue is legislation they brought in at the fishers' request to protect their fleets from nomadic scallop fishing. However, the IFCA are criticised for their reactive way of working because it allows them to be influenced by the best organised and most vocal and engaged communities. For example, a shellfish stock crisis divided the community, in that a powerful and vocal group wanted escape hatches whilst the rest of the district favoured a pot limitation scheme, and the IFCA favoured the escape hatches proposal. The IFCA has changed its consultation method away from the traditional model of open meetings because these are highly confrontational, and instead has a system of running consultations on a one-to-one basis, whereby a fisher has a private meeting with two officers. However, this change is not welcomed by all fishers: one influential fisheries association argues that fishers from their community are unlikely to take up allocated slots as they feel intimidated by being alone with two uniformed fishery officers. But other communities are unhappy at the power exerted by that association at open meetings, and blame the IFCA for not addressing this power imbalance. So IFCA Nine is seen as well-intentioned but weak in its attempt to engage all communities, and as a result, many fishers have reverted to passive resilience.

IFCA Seven claims they are committed to help fishers adapt their techniques so that the management impacts of marine parks would be as light as possible. They are also working with the Marine Stewardship Council (MSC) to support their fisheries in maintaining accreditation which enhances their marketing, and they consulted fishers at length before introducing their whelk management strategy based on a permitting scheme, thereby avoiding having to impose draconian regulations. The IFCA also acceded to the fishers' vehement protest against the proposal that their bad weather fishing grounds be closed. A fishermens' association (FA) lead says *'We have an IFCA that is quite pro-fishermen in our area'* (KI-64). However, fishers are deeply opposed to the IFCA's vessel monitoring system (VMS) program, viewing it as a 'big brother' measure, and do not trust the IFCA to use it to help make their management less precautionary. So it is a patchy picture in IFCA Seven—facilitating adaptive resilience strategies in some respects but not all.

The third group of IFCAs, the inactivists, are accused by respondents of moving away from meaningful engagement with fishers because of pressure of other work and reduced resources. For example, minimal communication is made with fishers: typically, information is uploaded on to the IFCA website or an email is sent out to FA leads, and the onus is very much on the fishers to check the website to discover policy changes, and on the FA leads to disseminate information. This inactive communication strategy mean that some fishers are unaware of management decisions until they are finalised. IFCA Six explains that marine conservation has become their primary objective. Members of one community in IFCA Six's district claim they have long tried to contribute to the work of IFCA but their efforts have been thwarted by the removal of one of their members from the IFCA committee, which fishers believe was due to MMO objections. IFCA Six gives local fishers little encouragement to develop adaptive, let alone transformative, strategies, as illustrated by a campaign to put pot limitations in place which fell on deaf ears, and many fishers have chosen passive resilience as their default strategy. However, in Defra's review of IFCA Six, relations with stakeholders were deemed to be improving: "Communication and engagement with stakeholders was perceived to be an initial weakness but effort has been committed and noticed over the establishment period...Development of the Community Voice engagement method was noticed and received positive comment" (Defra 2015, p. 52).

IFCA Four recognises that the survival of the fleet is integrally linked with its ability to adapt, and officers attempt to facilitate adaptive resilience by building up VMS data to understand where the fishing fleet work so they can refine their management tools to avoid indiscriminate measures. Furthermore, the IFCA tries to keep the cost of permits as low as possible and stretches each permit over a couple of years. However, the IFCA's main focus is on marine protected area (MPA) management and on what fishers must do to protect fish stocks, and fishers claim that membership of the IFCA committee is tilted towards conservation in a way which means that the fleet's interests will always be outvoted. Fishers criticise both the lack of physical visits by IFCA officers and the poor electronic dissemination of information by the IFCA whereby fishers are directed via their FA to a website, complaining that sometimes they are not informed about, let alone consulted on, important regulations, such as bans on netting and changes in lobster landing sizes, which leaves them inadvertently liable for prosecution. IFCA Four's conception of adaptive management seems to have more to do with adding conservation areas than with facilitating fisher's adaptive resilience strategies: *'as part of the adaptive approach it means that as new information comes along, we can be flexible in our management so if we find there are new areas around the European marine sites... which need protecting then we can change the boundary lines fairly easily'* (KI-105).

However, in Defra's review of IFCA Four, stakeholder involvement was praised: "Engagement is perceived to be good. Effort is committed to communication" (Defra 2015, p. 25).

IFCA Eight say they are accountable not to the fishermen operating in their district but to the requirements of the Marine and Coastal Access Act (MCAA) of 2009, which set up the IFCAs with a clear set of environmental objectives which they are obliged to meet. One of these objectives is to ensure the stocks are in the best possible state, and to present fishers with fishing opportunities that are consistent with that priority, and it is up to the fishers to decide how to capitalise on those opportunities by adapting their fishing activities. This is not to deny that IFCA Eight sometimes collaborates with fishers: one such collaboration is to facilitate a co-management strategy for the cockle and mussel fishery. The IFCA is willing to engage in further collaborations, but insists that the initiative has to come from fishers. The IFCA sees fishers as their own worst enemies and is irritated by their self-conferred victim status and unprofessional and impolite manner in communicating with the authority. IFCA Eight has reduced its communications strategy to an online platform, although it is willing to set up ad hoc meetings on request, so long as fishers take part in a professional manner. For their part, fishers are angry that the IFCA fails to take action when they provide evidence that justifies it—such as for a pot limitation scheme. In Defra's review of IFCA Eight, relations with stakeholders were favourably judged: "The IFCA's approach and commitment to engagement with stakeholders was acknowledged to be good" (Defra 2015, p. 29).

IFCA One holds a very pessimistic view of the future of the inshore sector. It believes it is witnessing the demise of the fleet as a result of declining stocks, legislation which leaves fishers with few adaptive options, and fishers' resistance to embrace environmental protection measures. The IFCA pursues a top-down management strategy in which consultation processes consider managerial ideas that have already been agreed upon. Their communications strategy is an entirely inert one, whereby they upload information to their website and rely on FA leads to disseminate this link and its contents out to their members. They are willing to set up ad hoc meeting upon request but would not tolerate any unprofessional behaviour from their fleet members within these forums. Some of the IFCA's decisions are seen by fishers as unfairly heavy-handed. For example, it imposes a charge of £500 pa for a cockle licence, which outrages hand pickers who object to having to pay anything for access to their traditional grounds. Also it granted a dredger a permit to collect mussel spat from the middle of the fishers' traditional grounds, leaving their fishery void for the following winter. This decision was based on theoretical modelling rather than empirical evidence, and fishers' protests fell on deaf ears because the IFCA declined to respond to their objections. However, an IFCA officer says they often follow fishers' initiatives: '*we don't usually have to go looking for stuff to do. They usually come and tell us when they want things to happen*' (KI-110. In Defra's review of IFCA One, stakeholders' comments were mixed: "Members of IFCA staff were commended for being hard working, open to meetings with stakeholders, responsive and described as an asset to the Authority…[but] Respondents noted the IFCA could make greater efforts to engage with stakeholders" (Defra 2015, pp. 42, 43).

Summing up this section, it is clear that IFCAs vary in their responses to the resilience strategies adopted by inshore fishers. Some IFCAs go out of their way to encourage and support adaptive and (to a lesser extent) transformative resilience strategies; some IFCAs are not prepared to initiate but only to respond to fishers' initiatives; and some IFCAs are reluctant even to respond to such initiatives.

6.4 Can Different Resilience Strategies Co-Exist?

The question here is whether fishers and/or communities can adopt more than one resilience strategy simultaneously. The answer to this question is 'Yes'. It is evident that passive, adaptive, and/or transformative strategies can occur simultaneously in a community: indeed, in most communities we found different resilience strategies operating at the same time. For example, in three communities—North Eastern B, Kent and Essex A and Sussex A—transformative resilience is combined with adaptive resilience; in one community—Cornwall B—adaptive resilience is combined with passive resilience; whilst in ten communities—Southern A, Devon and Severn B, North Western B, Kent and Essex B, Northumberland B, Eastern B, North Eastern C, Eastern A, North Eastern A and Northumberland A—there is evidence to a greater or lesser extent of all three resilience strategies. The remaining communities are omitted as they exhibit only one resilience strategy.

Of the three communities adopting both transformative and adaptive resilience strategies, North Eastern B stands out for its transformative initiatives. The community has a young, united fleet with an extremely energised, trusted and well-connected cooperative, boosted by strong historical and family connections, able to represent the fleet and promote the industry. This community manages to transform threats into opportunities, as illustrated by fishers' success in challenging a large offshore wind-farm development proposal sited over their best ground. The FA won substantial compensation money and ensured that their most valuable fishing area was not affected, and the compensation money was used to create a strong and profitable business selling fuel to all harbour users, which enabled it to employ a lead organiser/consultant and purchase a marine research vessel. This association now exerts significant influence over the IFCA: for instance, inshore fishers were concerned about offshore scalloper vessels destroying their ground so they pressured the IFCA to ban them from inshore waters. At the same time, the fishers have the legal flexibility to adopt adaptive strategies to take advantage of all the opportunities that present themselves. For example, they exploit an intermittent and brief fishery presented during the annual migration of the velvet crab (*Necora puber*). Their united and effective FA enables them to successfully pursue other opportunities. For instance, they secured FLAG funding to promote the fleet locally through food festivals; they started a v-notch project to prove how sustainable they are; they used their seat on the IFCA to overturn a pot limitation proposal; and they networked with strong local and national links to the local council, IFCA and the NFFO.

In Kent and Essex A, there is a small but extremely tight-knit (again based on family connections) fishing community led by a strong and active fishing association, which is closely linked to both NUTFA and neighbouring communities. The association has engaged in three transformative initiatives, the first of which was Hugh Fearnley-Whittingstall's 'Fish Fight' project as a mechanism to fight the quota system. Unexpectedly, however, this campaign turned into the no-discard policy of the EU's landing obligation, which posed a great risk to the local fishers' survival since they were highly reliant on quota species and worked in a mixed fishery. So they switched the focus of their transformative campaigning to Brexit, closely aligning themselves with Nigel Farage's UKIP and taking part in the Thames regatta protest. Their third transformative initiative was to fight the MCZ process, liaising with other affected communities to launch a challenge to one of the sites taken forward as part of the government's roll- out of MCZs in English inshore waters. They were initially successful in that the proposed MCZ site in their area did not go ahead, but it has now been implemented by Defra in a revised form. At the same time, local fishers continue to fish in an adaptive fashion within the constraints that the legislation currently allows them, diversifying where possible.

In Sussex A, there is a fiercely united fishing community led by an influential, innovative and dynamic association which works closely both horizontally with other fishing communities and local institutions (including the university and the local council), and vertically with national NGOs such as Greenpeace, the New Economics Foundation (NEF) and NUTFA. Their FA has an impressive record of fighting for transformative changes. For example, they improved pricing and market access by establishing their own market run by their fishing community; they introduced a levy from which fishers can be paid poor weather payments; they attempted to transform their quota allocation crisis by getting actively involved with Hugh Fearnley-Wittingstall's 'Fish Fight' project, though like other involved communities got badly burnt with the resulting no discard policy; they linked with NUTFA and other communities together with Greenpeace to take Defra through a judicial review to bring all quota back into public ownership and allow it to be reallocated in a fairer fashion; they met the threat from trawlers in their prime shell fishing ground by pushing hard for an MCZ and by leading the research required; and they pressed for an inshore PO to enable them to take full control of their own quota allocation. They also have an adaptive strategy: for example, they won FLAG funding to improve the working conditions of their fishing community, and they diversified their fleet, taking advantage of emerging fisheries such as a returning herring fishery, changing their operations to target it and contributing to a local food festival to promote the marketing of herring.

In Cornwall B, adaptive resilience is combined with passive resilience. Whilst this community is united, they appear somewhat detached from their association which rarely calls meetings. The FA is also not connected with either neighbouring communities or external groups, and works independently. Fishers exhibit adaptive resilience, whereby legal flexibility and fortunate positioning enable the fleet to pursue different types of fisheries to avoid regulations such as tight quota restrictions, and they are all fairly profitable. However, passive resilience is creeping in: many fishers say they feel powerless in relation to the MMO and make allegations of insincere consultations and hostile and malicious treatment of some community members. It seems likely that the next time the community's fishers are faced with an external stressor that they cannot adapt around, more of them will fall into a passive resilience mode.

The ten communities where there is evidence of all three resilience strategies will now be discussed. In the Southern A community, the FA has mounted several transformative campaigns on behalf of the community. For example, in the face of restrictive quota allocation rules they fought for more aquaculture opportunities to be rolled out in their bay; to reduce violent and gang-related non-compliance they helped develop a by-law to manage the wild shellfishery; to protect the fisheries from nomadic fleets they demanded a pot limitation regulation; to improve access to their harbour they campaigned for better facilities; to deal with the huge challenge of the landing obligation they actively engaged with Defra to find a way forward for the community; and together with the IFCA they won FLAG funding. These initiatives benefitted from the efforts of an energised, enthusiastic and 'bolshie' leadership, made up of a chair, vice-chair and secretary, who reply to consultations and represent fishers at meetings. Also important is the organisation's policy of linking up not only with the IFCA but also with the local MMO, council and harbour boards as well as national groups such as NUTFA and NGOs such as Greenpeace and NEF. However, the FA are faced with a disgruntled fleet in which local conflicts undermine community unity. Very few of their fishers attend FA meetings, and many of them keep their heads down, preferring to work in isolation (passive resilience). Whilst disengaging with their FA, fishers still continue to work in an adaptive manner, targeting both quota fish and shellfish, and seeking alternative livelihoods outside the sector (such as building work) to keep them going when the fishery becomes unprofitable.

In Devon and Severn B community, four transformative strategies stand out. The first was to deal with poor market access and pricing: the fishers decided to market their fish themselves and built a new fish market with all the fishers becoming shareholders in the company, electing one fisher to become a paid full-time staff member heading up the company. The second transformative initiative was opposition to a proposed MCZ: the FA collaborated with NUTFA to provide evidence supporting the case for a reduction in the closed area. Third, the FA put pressure on Defra to acknowledge the challenge they faced from the discard ban and to amend the landing obligation regulation to reduce the risk to the fleet. Fourth, many fishers supported the Brexit campaign. However, despite the fact that the FA in this community is strong, united at a grass-roots level and enjoys close relations with adjacent communities and national organizations such as NUTFA and NFFO, it struggles to relate to the IFCA and MMO: no one from the community holds a seat on the IFCA committee, and fishers complain that their responses to MMO consultations have little effect. As a result, many fishers reported becoming disengaged from the management system, and, losing trust in authority figures, they concentrate on day-to-day adaptive strategies (including prosecuting new species such as spider crabs (*Maja squinado*), single-handed work, and black fishing), or resign themselves to passive resilience.

In North Western B community, the FA has a successful track record of transformative campaigns. For example, when dealing with a quota allocation crisis, their association contacted the MMO and proposed a community quota scheme whereby they would take control of managing their quota. When faced with poor marketing prospects, they used the money from a wind-farm community scheme to buy an ice-making machine and a chiller to ensure that the landed product was kept in excellent condition, and they bought a van to take the landed fish to markets which paid higher prices. When experiencing high landing costs and escalating fuel prices, they used money from the FLAG to establish a fuel business selling diesel to all the vessels using the port. When confronted by expensive and remote locations for mandatory training courses (e.g. on sea survival), they used some of the land they had purchased with the FLAG money to set up a training school, offering cut price courses to local members. To deal with the increased paperwork these initiatives generated, the association funded their leader on a full-time basis. One reason for the success of these transformative initiatives was the extensive networking carried out by the FA: it linked up with neighbouring communities, local management groups such as the IFCA, MMO and harbour board, and national union bodies in the shape of NFFO and NUTFA. However, whilst the association pursued transformative strategies, many fishers felt excluded from its decision-making processes because there were very few open meetings, and others were unhappy with its perceived deference to the IFCA and MMO. Behaviour characteristic of passive resilience emerged amongst many fishers, some of them choosing strategies of (destructive) adaptive resilience such as the reduction of crew and occasional black fishing.

In the Kent and Essex B community, the FA launched several transformative campaigns. For example, when faced with the threat of losing valuable fishing grounds because of the proposed siting of a telecommunications cable, they negotiated with the developers to improve their disruption payments and used their compensation money to overhaul the spiralling costs of diesel by establishing a fuel company owned by the fishers and from which they draw down the profits. To deal with the quota allocation crisis, they engaged with neighbouring communities to take a community quota project forward, and this morphed into a campaign to establish a national inshore PO and to join with NUTFA and Greenpeace to pursue a judicial review questioning the legality of quota ownership by the existing POs. However, although the inshore fishers won a partial victory in the judicial review (on the distribution of new or unused quota), they lost on another part (on the redistribution of existing quota holdings), and Defra declined to appeal on this issue. Moreover, although the FA enjoys community support and has strong local links with the council, MPs, harbour board and IFCA, as well as national links with NUTFA, UKIP and NGOs, several fishers expressed frustration with the negativity of the authorities, especially the MMO, whom they accused of entrapment of fishers in court cases. As a consequence, some fishers report disengaging with management, and moving into a passive mode. Other fishers, however, continue to pursue adaptive fishing operations, seeking opportunities wherever they can within the rules, targeting different fin-fish depending on quota limits and moving over to shellfish in the form of whelks, crabs and lobsters when quotas become tight, though they report that their ability to switch between vessels, gear type and fisheries is becoming harder because of tighter legislation.

In Northumberland B community, evidence of transformative resilience includes the community's concerted effort to overturn the IFCA's vulnerability assessment of damage caused by fishing gear to protected MCZ benthic features, which threatened to justify the introduction of highly restrictive gear management measures. However, fishers report that despite remaining united and passionate, their argument was completely ignored by their IFCA. As a result, some fishers voice passive resilience sentiments such as lack of trust in their IFCA and pessimism about being able to influence management. Most fishers, however, feel they are still able to adapt and function within the constraints they faced, though some admit that this is only possible by reducing crew, and (occasionally) by rule-breaking.

In Eastern B community, there is an aging and unprofitable fleet where younger crew have been lured away by the secure and lucrative work provided by an offshore wind-farm. The remaining fishers have collectively united to pursue several transformative initiatives. One involved action, supported by the local market and their MP, to fight the unfair quota allocation system, but they failed to persuade the MMO. Another involved engagement with the marine conservation zone (MCZ) project to ensure that their fishing ground was left open for fishing, but the fishers claim that the data they submitted to the project were used against them and as a result they were banned from their old fishing ground. Disillusion with these failures has led some fishers into a passive resilience mode, and they do not engage with management any more, nor with their FA, which has now dissolved. These fishers explain they still fish but only because they do not know what else to do. Other fishers say they have adopted a strategy of adaptive resilience. For example, those who still have crew say they have had to dismiss them and work solo due to financial problems. Others have moved to long-lining, though with no intention of using its light environmental footprint to apply for MSC accreditation.

In North Eastern C community, although there is an elected, active, determined and innovative FA, fishers report that it has not invested sufficient effort to develop an external support network, and they blame this deficiency for the failure of the FA's attempted transformative strategies. For example, for years, the FA lead and those close to him campaigned for a pot limitation to reduce effort on their shellfishery (as a result of other fishers adapting to avoid quota species), but they claim that because they were not supported by any external organisation, their pleas were ignored by the IFCA. Similarly, the association fought to win FLAG funding as a way to enhance their poor marketing strategy, but the money was allocated to other communities as they lacked external cheerleaders. This led some fishers to lose faith in their FA and move into a passive resilience state, discouraged by increased predation from seals, tight quotas and closed areas. Others pursue adaptive strategies, but of a risky or destructive kind, for example, by shedding crew or illegal fishing.

In Eastern A community, passive resilience can be seen in its aging fleet which feels powerless to protect itself from policies which have reduced fishing opportunities. Many continue to fish because they see it as part of their identity, but know it is no longer a profitable or healthy sector. Others try to transform their fishery situation: for example, their FA played an active part in the group which secured FLAG funding for the area. However, out of the large amount of funding won, they were only allocated expenditure for one second-hand tractor to help them move their vessels up the beach, while non-fishing groups were given thousands of pounds. The FA tried another transformative strategy, presenting a paper to their IFCA recommending new policies to achieve conservation objectives whilst offering protection to the inshore fleet. But this initiative got no support from the IFCA, the local council, MMO, NFFO or NUTFA, and many local fishers say they are increasingly disillusioned and mistrustful of the establishment, and are shifting towards passive resilience behaviour. Adaptive resilience is also evident in skippers choosing solo fishing as it becomes uneconomic to employ crew, despite their awareness that this is an unsafe practice.

In North Eastern A, transformational moves include attempts by the FA to get the IFCA to impose a lobster pot limitation to combat spiralling effort as vessels move into this fishery and away from quota fish, flooding the market and threatening the lobster stock's sustainability. The IFCA ignored their proposed strategy and opted instead for an escape-hatch panel system. The fleet explained that sizeable males are often long and thin and can easily extricate themselves through the escape hatch, thereby losing shell-fishers up to 30% of their catch, but the IFCA went ahead regardless. Many fishers left their FA, saying they see no point in continuing to fight what they perceive to be a corrupt administration, and adopted mixture of unsustainable adaptive strategies and passive resilience. The FA itself persisted with its transformative strategy, seeking new ways of persuading the IFCA to reverse its decision, including getting the media involved, and asking NUTFA to help.

In Northumberland A, a rather similar picture emerges, in that the fishers have a FA leader who works very hard to achieve transformative strategies. For example, he used his position on the IFCA committee to ensure that the new tranche of MCZs were not taken forward in an extremely precautionary fashion which would further restrict fishers' target grounds, and he constantly responds to other development consultations which threaten similar impacts. However, he faces considerable difficulties: he carries out this role whilst operating as a full-time fisher which places him at high risk of burn-out; the number of issues needing his attention is becoming overwhelming; he has to deal with an unhelpful IFCA committee; and he is given no support from his community. He has not developed external networks with organisations such as the NFFO or NUTFA, so in effect he is engaged in a one-man battle. Many fishers have adopted a passive resilience stance, believing they have to accept whatever is thrown at them, feeling trapped in their shell-fishery because they are losing their fishing grounds to protected areas and offshore developments, and they are deeply suspicious of the way their FA engages with the IFCA. Other fishers follow unsustainable adaptive strategies such as reducing crew and black-fishing.

Clearly, many permutations and combinations of resilience strategies can exist within a community. This conclusion underlines the assertion by Béné et al. (2014, p. 614) that we should see the three resilience strategies as complementary not mutually exclusive, still less competitive: we "need to move away from the either/or discourse adopted in the literature, where the three dimensions of resilience (absorptive capacity, adaptability and transformability) are presented as antagonistic and excluding each other" (see also Béné et al. 2012). Moreover, a single fisher may exhibit passive, adaptive and transformative tendencies at different stages of his/her career, and this fluidity is the subject of the next section.

6.5 Are There Any Trends in Resilience Strategies?

This question raises the issue of whether there are patterns of change in resilience strategies adopted by the English SSF. Resilience strategies are not static—indeed, most of them constantly change over time—but do they change in predictable or in random ways? Only in one community—Southern B—has a single strategy (adaptive resilience) been followed throughout its recent history. The reason why this community is so consistent is partly because of a fortuitous conjunction of favourable fisheries opportunities (they can alternate between quota species, switching between trawling, hand-lining, and shell-fishing); partly because none of their fishing grounds has been designated as an MCZ site; partly because they have an active, supported and inclusive FA; partly because their FA lead has created a good relationship with both the IFCA and the MMO; and partly because the community has remained close and united.

Both Island A and Devon and Severn A communities have settled back into a strategy of adaptive resilience following a period of transformative resilience. In Island A community, a campaign of transformative resilience created a relationship with their IFCA in which it became largely self-managed. The IFCA now bring their potential policy ideas to the community, who have the final say on whether or not the ideas are taken forward. For example, the IFCA proposed a lobster pot limitation scheme which the fishers rejected as unnecessary. They were able to exert this veto power because they had worked hard in the past to prove that their fleet and the stock they target were sustainable through lobster tagging, working with local divers who collect underwater evidence through video and photography, a lobster v-notching project for stock control purposes, extending carapace landing sizes for lobsters, imposing a closed winter season, banning the sand eel fishery to support the sea bird colony, and establishing a hatchery for crawfish to help the population recover faster. All these transformative measures contributed to building a platform on which an adaptive strategy could be established, and at the time of research, the Island A community fishers have gone back to working adaptively by taking advantage of other fisheries when they become available such as the crayfish and ray; testing new lobster grounds; pursuing alternative livelihoods; trying different marketing options such as selling directly to tourists and local restaurants during the short tourist season; and cutting costs by employing family members to pack the crabs instead of sending them to a processing plant. It is clear that the fishers can switch back into transformative mode if needed, but at present they are hugely supportive of their IFCA and extremely optimistic about their future.

In the community of Devon and Severn A, the fishers have found ways to force through several transformative solutions. For example, facing a potentially lethal EU policy in the form of a total drift netting ban as a result of turtle by-catch in the Mediterranean, which would have ended their herring fishery, they networked with the NFFO to successfully negotiate a derogation for UK fishers. Another example is their collaboration with neighbouring communities to impose a four-year voluntary closure on a struggling target fishery which allowed the stock to recover. Also, they used their excellent community and council connections to win FLAG funding for their area, and to organise fishing festivals celebrating their fleet which helped them market their stocks. Having won these campaigns, they have reverted back to working in an adaptive fashion: taking up alternative work outside the fleet such as building and decorating work; investigating sail power to supplement engine power; taking advantage of positive pricing; and targeting highly sustainable and healthy stocks of shellfish and herring using traditional low-cost methods. They are fortuitously located in a sheltered bay so have a competitive edge over neighbouring communities, but they are well aware of the limitations of being a tiny fleet in Devon, so they maintain close links with the NFFO, holding a position on their committee to ensure that when needed, their case will receive support.

By contrast, in two other communities—North Western A and Sussex B—there has been a shift from transformative to passive resilience. North Western A's FA (vainly) attempted to transform an IFCA policy which threatened the viability of their hand-gathering operations by permitting a large industrial fishery to access shellfish spat they traditionally harvested. This failure was compounded by the shift from the SFC to the IFCA system of governance, whereby committee members were no longer representatives of their communities but instead deemed as experts in their field. As a result, the three fishers who were IFCA committee members removed themselves from the committee IFCA and reverted to a passive fatalist mode, explaining they were no longer able to diversify their fishery any further due to a mixture of age, lack of education, and permitting schemes which reduced their room for adaptability.

In Sussex B community, the fleet attempted transformative strategies by fighting for their IFCA to impose a whelk pot limitation to deal with the increased pressure on this fishery as people left the quota fishery. However, their attempts were unsuccessful and they feel they have been ignored. They tried one further transformative move—to prevent a large offshore wind farm development being positioned over their most profitable fishery—but again failed, and this development cut them off from their prime fishing areas. Respondents say that this setback has destroyed their energy to fight, and they have come to the conclusion that they are powerless over the system that regulates their work. They are frustrated and demoralised and their FA has been dissolved, ending any further chances of transformative initiatives. They are isolated from neighbouring communities and NUTFA as well as the

IFCA, and they have withdrawn from engagement with the MMO. They say they were unable to adapt their fishery to deal with restrictive quota legislation which has made it impossible to compete with larger vessels. So they have become passive resilients.

In one community—Cornwall A—there has been a shift from adaptive to passive resilience. This community has seen its FA decline into inactivity as its fleet is aging and there are no new recruits signed up to join it. Respondents explain they have tried to adapt their fishing operations to circumvent the increasing restrictions imposed on them which have removed opportunities to the point of fencing them into a single form of fishing—crab and lobster. They explain that when fishing quotas tightened, they moved temporarily into the crab and lobster fishery, but this move was used by the MMO to justify capping their licences, so they lost the ability to move between this fishery and targeting quota'ed fish. The community has some transformative ideas, including establishing their own shop through a cooperative, but they lack confidence to follow such ideas through. They have seen the way that transformative projects such as Hugh Fearnley-Whittingstall's 'Fish Fight' have been manipulated by national and EU systems and used against them. The result is that many fishers are pessimistic about the future of their fishery, and have embraced passive resilience.

In summary, although there is no inevitable or uniform direction of travel—each community is unique—we can make five generalisations about the factors which determine patterns of change in resilience strategies. First, where there is a united community, with a strong, inclusive and transparent FA, and a vigorous local leadership who have developed a supportive external network, the odds are that the community will not decline into passive resilience. Second, managerial responses are closely linked to resilience strategies, and the most common reason for a community's decline from transformative resilience to passive resilience is disillusion by fishers with the perceived negative response of some IFCAs and the MMO to their transformative initiatives. Third, it is common for fishers and their wider communities to carry on adaptive strategies until they become untenable, and at this stage, they attempt a transformative change, which either succeeds and enables them to return to the adaptive mode, or it fails and they fall into the passive mode. The fourth generalisation is that the capacity of a community to take on challenges through transformative strategies is finite: challenges may become so overwhelming that communities realise that even herculean efforts are not enough to achieve a positive and sustainable outcome and they will revert to passive resilience. Fifth, in many areas passive resilience has become the dominant mode and many English fishing communities are losing the battle to make their inshore fleets thrive. The transformative efforts of fishers' associations in these areas have been unsuccessful in creating a platform for a long-term adaptive strategy to be established, and more and more fishers are either leaving the industry or adopting a stance of passive resilience, or pursuing a short-term strategy of unsustainable adaptive resilience. This is a profoundly worrying situation because it threatens to eliminate the English inshore fleet or reduce it to the status of a backwater activity in many areas. Once a community has declined into passive resilience, it finds it very hard to emerge from that condition: there were no examples of communities shifting from passive resilience to adaptive or transformative resilience. When passive resilience takes hold in a fleet and its association, it can spread to destroy unity and any capacity to fight. Once a belief that they can make a difference is lost, it is very difficult for fishers and communities to regain confidence. It seems that unless external intervention takes place, a spiral of negativity will reinforce itself until the fishing community itself is at risk. To avoid such a tragedy, the authorities (especially the IFCAs) need to examine what are the factors causing this crisis in the inshore sector in their districts, and investigate what measures can be taken to avert it. One suggestion is that they might re-examine their engagement and communication mechanisms.

6.6 Are Some Resilience Strategies More Sustainable than Others?

In answer to this question, transformative resilience cannot be sustained indefinitely because it is too heavily resource-intensive and risky for a community. It requires great bursts of energy and is the costliest of all the resilience options, and the viability of a community being able to use this as a strategy depends on how previous attempts have been received and what governance arrangements are in place. It is unlikely that passive resilience could be a viable long-term strategy for many fishers, since they would feel defenceless against increasing restrictions on inshore fishing, and eventually they might have to give up fishing altogether. Adaptive resilience is the most sustainable strategy in the long-term, because fishers can go on adapting to changing conditions endlessly provided that the circumstances under which they operate are favourable. It could be argued that the role of a transformative strategy is to ensure that those circumstances remain favourable over time. In other words, transformative resilience is a necessary condition for achieving a sustainable future for inshore fishers by building a platform for their fishing activity that can be pursued adaptively for generations to come.

6.7 Conclusion

This chapter has discussed five issues that arise from chapter five on the resilience strategies adopted by English inshore fishers in response to the challenges facing them that were outlined in chapter four. The first issue is whether passive, adaptive and transformative resilients have different psychological profiles, and we found they do. The second issue is whether Inshore Fisheries Conservation Authorities (IFCAs) facilitate the resilience strategies adopted by fishers in their districts, and we found that some IFCAs are more facilitative than others. The third issue is whether two or more resilience strategies could exist simultaneously in a community, and we found that most communities have more than one strategy, and many have all three strategies being pursued at the same time. The fourth issue is whether there are trends or patterns of change in the adoption of resilience strategies, and we found that successful transformative resilience strategies often pave the way for adaptive strategies to be adopted. The fifth issue is whether some resilient strategies are more sustainable than others, and we found that adaptive resilience is the most sustainable strategy. This discussion has shown us that resilience theory enables us to understand the problems facing the English small-scale fishery (SSF) and to offer some prospects of solving those problems (Symes et al. 2015). The final chapter discusses the question of what kind of future lies ahead for the English SFF.

Chapter 7
Conclusion

> *'Small-scale fisheries are not a problem to be solved, but a solution to be unfolded' (Fábio Hazin)*
> *'Small-scale fisheries are always seen as a problem, rather than a viable part of the solution of the crisis of fisheries' (Denis Pauly)*
> *'small-scale fisheries are now in focus as never before' (Svein Jentoft and Maarten Bavinck)*

7.1 Introduction

This book analyses the current marginalised state of the English small-scale commercial fleet. The last 40 years have been a period of significant change for the English fishing industry with the inshore or under-10 m fleet disproportionately adversely affected. As a result, it has been predicted that the long-term viability of many of these fishing communities is under threat (Defra 2009). This could lead to an economic, social and cultural disaster for many isolated and deprived coastal fisheries communities in which they are embedded (Urquhart et al. 2011). Much has been written about the causes of vulnerability, pointing especially to a lack of quota, which until recently has been the main topic discussed in the trade publications, and major campaigns have highlighted the inability of fishers to make a living from their quota allocations. However, the central finding of this book is that lack of quota is a symptom and not the root cause of the problems faced by inshore fishers. The fundamental cause is lack of fisher participation in fisheries management. As Chuenpagdee and Jentoft (2015, p. 14) put it, "The heart of small-scale fisheries concerns is governance".

This concluding chapter addresses the implicit question in the research: 'Have English SSFs got a future?' and answers 'Yes, if certain conditions are met'. For some respondents, that future will be secured by Brexit which they expect to provide such a large increase in quota for the UK that it will transform the prospects of

© The Editor(s) (if applicable) and The Author(s), under exclusive license to Springer Nature Switzerland AG 2021

R. Korda et al., *Resilience in the English Small-Scale Fishery*,
https://doi.org/10.1007/978-3-030-54245-0_7

the English SSF sector. We do not share this expectation, since the Brexit outcome for UK fisheries is likely to be complex and uncertain, and the English SSF sector may lose more than it gains from leaving the Common Fisheries Policy (CFP). In our view, the main condition that must be met for the English SSF sector to have a future is giving inshore fishers a greater degree of participation in decisions which affect them. It is not a gift from above that the English SSF should be looking for, but a governmental recognition of their capacity to co-manage themselves.

7.2 Brexit Is Not the Panacea

Brexit can be interpreted as an affirmation of transformative resilience by restoring sovereignty to English fishers, and it seems a rational step to take to overcome the quota crisis for those communities who target quota'ed fish in a mixed fishery. It is thereby unsurprising that this strategy is particularly popular with fishers in the south-east and south west whose main issues are quota-related, exacerbated by the EU's landing obligation. Inshore fishers perceive that their future is never going to be rosy whilst the UK is subject to the CFP, governed by the European Commission. Accordingly, many inshore fishers have embraced the Brexit process, closely aligning themselves with Nigel Farage, and taking part in the Thames regatta protest. In 2016, 92% of UK fishermen voted in the referendum on the UK's continued membership of the EU in favour of leaving the EU, primarily because they believed that the CFP has caused the decline of the UK fishing industry in recent decades. Most UK fishers have high expectations of a Brexit dividend that will uplift fishing opportunities for the UK industry by overhauling the CFP's historic access agreement based on the principle of 'relative stability' (Phillipson and Symes 2018). The publicity surrounding Brexit has moved the small-scale fleet into the limelight and given them an unprecedented national platform drawing support from NGOs, MPs and the general public who are determined to use Brexit to preserve fishing in coastal communities not least by reducing quota inequality between the inshore and offshore fishing sectors. In few other areas of UK public policy does the idea of 'taking back control' assume such a literal meaning.

However, there is considerable uncertainty as to how Brexit negotiations on the fisheries question will unfold. Whilst British negotiators draw up detailed plans for a post-Brexit future to grant the UK fishing industry's aspiration for significantly increased fishing opportunities, European institutions are, in parallel, demonstrating their determination to protect existing access rights and quota entitlements for Europe's fishermen (Phillipson and Symes 2018). It is possible that the UK may not emerge with the radical changes hoped for by the majority of UK fishermen: indeed, at the present moment, it is not clear that the English SSF will be better off post-Brexit. One authoritative source, the New Economics Foundation (NEF), conducted a scenario analysis of six possible Brexit outcomes—no Brexit; hard Brexit; soft Brexit; fisheries-first Brexit; fisheries-last Brexit; and no-deal Brexit—and concluded that on all but the fisheries-first scenario, "Brexit will not improve the economic performance of many small-scale fishers that are currently struggling" (Carpenter 2017, p. 5). One worry for the inshore fleet is whether the UK government, even if it secures increased fish quotas from Brexit, will be more generous to the inshore fleet post-Brexit than it has been pre-Brexit. Another worry is that after Brexit, not all foreign vessels in English waters would be automatically excluded from fishing quotas, since some of them are registered and flagged in the UK and have purchased quotas from UK fishers. A third worry is that loss of access to the single market, now a clear policy of the hard Brexit lobby, could undermine the marketing prospects of some English inshore fleets, especially those dependent on exporting shellfish products to France and Spain. Ironically for the English SSF, as Britain completes the process of leaving the EU, the CFP is introducing three measures to help SSF: giving Member States the right to restrict fishing within 12 nautical miles of their coast lines; excluding SSF from individual transferable quota (ITQ) schemes; and conferring financial benefits on coastal fishing communities (Lloret et al. 2018). These measures emanate from the sentiments expressed by the European Commission in its Green Paper (EU Commission 2009) proposing a differential approach to the management of large-scale and small-scale fisheries sectors in Europe: the large-scale sector would remain governed centrally as an economic unit with effort reduction and individual tradeable quotas; whereas the small-scale sector would be managed locally as a social unit allocated non-tradeable community quota (Symes 2014). Also, as Davies et al. (2018, p. 209) note, Article 18 of the European Maritime and Fisheries Fund (EMFF) Regulation requires Member States which have more than 1000 SSF vessels to produce "an action plan for the development, competitiveness and sustainability of small-scale coastal fishing".

7.3 Participation in Fisheries Management Decision-Making Is Essential

The message that comes out of our research loud and clear is that the most fundamental demand of English SSF, and the only way of securing a sustainable future for the English SSF, is to increase its meaningful participation in inshore fisheries decision-making. This is a message that confirms what the fisheries press say:

> "One of the biggest problems for fishermen is their disconnection from the people who manage their activities. They have complained for many years that the officials who manage their fisheries are remote and office-bound, and do not understand the complexities and practicalities of fishing. This is particularly the case in inshore fisheries, where local management is faced with a wide range of local factors that vary from region to region, and require specific management measures that can differ even within local management districts…There is a growing demand by fishermen to be more centrally involved in the management of their industry on which their livelihoods depend" (Linkie 2019b).

Lack of participation is a problem not only for English SSF, but also for SSF in Scotland: Pita et al. (2010, pp. 1099–1100) found that "Scottish inshore fishers… perceived themselves not to participate in the decision-making process". Ifremer

(2007, pp. 431, 434) reports on SSF in other EU Member states that "participation in national institutions is average or weak and almost non-existent at the Community level. As a result, low political power makes them extremely vulnerable to pressures arising externally especially from LSF and tourism", and notes the "need for participation, through greater and broader stakeholder involvement from conception to implementation of policy, at local and regional level". Lack of SSF participation in fisheries decision-making in other parts of the world is a problem flagged up in many published studies, including Sathyapalan and George (2015), Charles (2011a), Béné et al. (2007), Pita et al. (2015), and McConney and Charles (2012). Béné et al. (2015a, p. 14) write that "small-scale fisheries are still seriously under-represented in the governance bodies of most of the ocean initiatives driven by international institutions and organizations". Kurien (2011, p. ix) refers to the need for fishers "to take control of their livelihoods" and for "empowering small-scale fisheries and their communities". Andrew et al. (2007, p. 233) state that "A major impediment to resilient SSF in the developing world is the inability of fishers to secure and exercise rights and responsibilities over fisheries resources". Kosamu (2015, p. 366) points out that "Several scholars...have decried the lack of recognition and involvement of small-scale local fishers in fisheries management, especially in developing countries". Ratner and Allison (2012, p. 375) claim that

> "The most significant obstacle to realising the development potential of small-scale fisheries does not lie in fisheries science, technology or science, though there is work to be done in each of these areas. The emerging consensus in the international development community is that the most significant obstacle is governance...there is no single right answer to the appropriate balance of goals for fisheries development...Critical instead is the *process* of arriving at goals considered socially legitimate by relevant stakeholders" [italics in original].

Likewise, Cohen et al. (2019, p. 6) state that "retaining the benefits SSF provide to society, requires improved representation of SSF in international, national and multi-stakeholder policy and investment arenas". Charles (2011a, p. 288) reports that the World Forum of Fisher People (WFFP) and the International Collective in Support of Fishworkers (ICSF) have both declared that SSF have a human right "to participate in fisheries and coastal management decision-making". Jentoft and Chuenpagdee (2018, p. 312) agree: "small-scale fishing people can legitimately claim that participation is not just a functional mechanism, something that will help solving wicked governability problems, such as IUU (illegal, unreported and unregulated) fishing, but a fundamental value, a human right, related to social justice and human dignity, and hence wellbeing". This view reflects the priority of the human rights approach to SSF over the sustainable livelihoods approach to SSF. On this view, justice is prior to welfare: indeed, justice is a precondition of welfare. As Jentoft et al. (2011, p. 460) put it:

> "poverty alleviation, however well-intended and warm-hearted, cannot and should not be imposed on people from the top-down. It matters to people how things come about, whether a particular good is handed to them or created by them...This is also an issue of human dignity and freedom. Therefore, it is essential to encourage their involvement and facilitate their empowerment. Indeed, Knudsen argues that lack of capacity to participate can be considered a dimension of poverty".

Basurto et al. (2017, pp. 50, 51) say that in their research, "almost all answers to the question of what has been effective in supporting SSFs, as well as citations in many studies in the literature, describe efforts to return more control of SSFs to fishers and fishing communities…the question today is no longer focused on *whether* or not states should return or share more of the governance of SSF with the users, but *how*" [italics added]. FAO (2005, pp. v, 32, 71) declares about its vision for the future of SSF that

> "Ensuring greater participation by small-scale fishers and their communities in the formulation of policies, the development of related legislation and regulations, and in management decision-making and implementation processes, is vital to the realization of this vision… without participation by, and empowerment of, both small-scale stakeholders and civil society, success is likely to be limited…The biggest single contribution to achieving this goal probably lies in the empowerment of small-scale fishers and fish-workers within a context of transparent and open engagement".

FAO (2015, pp. 1, 3, 7, 17) states that the objectives in its guidelines for securing sustainable SSF

> "should be achieved through the promotion of a human rights-based approach, by empowering small-scale fishing communities…to participate in decision-making processes… ensuring active, free, effective, meaningful and informed participation of small-scale fishing communities….in the whole decision-making process related to fishery resources and areas where small-scale fisheries operate…States should involve small-scale fishing communities…in the design, planning, and, as appropriate, implementation of management measures…affecting their livelihood options. Participatory management systems, such as co-management, should be promoted in accordance with national law…States and other parties should enhance the capacity of small-scale fishing communities in order to enable them to participate in decision-making processes…through the creation of legitimate, democratic and representative structures".

Smith et al. (2019, p. 52) claim that meaningful SSF participation in fisheries management is the only way to resolve vexed problems like the allocation of quota: "small-scale fisheries are likely to have allocation goals that differ from industrial fisheries. These may include promoting equity, preserving cultural values, ensuring the participation of marginalized groups, or setting aside fish for subsistence, among others. A process for allocating rights will require extensive stakeholder participation to elucidate these underlying goals and values and to develop solutions that best address them". Salas et al. (2007, p. 14) claim "there is wide recognition that it is necessary to make the management process more participatory and transparent". Similarly, McGoodwin (1990, p. 189) asserts: "Because the most pressing need in the fisheries today is to reconceptualise management policies in a way that makes human concerns paramount, it is essential that future management regimes provide for more extensive participation by the people concerned. What is most needed now is a shift away from autocratic and paternalistic modes of management".

Msomphora (2015) reports from her study of two Scottish Inshore Fishery Groups (the Scottish equivalent of IFCAs in England) that there was a strong correlation between stakeholder participation in the management plans and stakeholder satisfaction with the management plans. Moreover, as Lowther and Rodwell (2013, p. 1) note, the very legislation that brought the IFCAs into being—the Marine and

Coastal Access Act 2009 (MCCA)—requires them to involve stakeholders in their decision-making processes: "The aim is for IFCAs to overcome the more limited scope of the predecessor structures, the Sea Fisheries Committees, and to contribute to a contemporary, open and inclusive governance model…The MCAA approach is to create better opportunities for stakeholder engagement". North Eastern IFCA (NEIFCA 2015/16, p. 6) state that one of the seven criteria that IFCAs nationally have agreed to adopt, is: "IFCAs work in partnership and are engaged with their stakeholders". For its part, the Marine Management Organisation (MMO) (2016b) claims to be committed to stakeholder participation by SSF. John Tuckett, then CEO of the MMO, said: "There is a whole agenda on how we engage with the U-10m fleet. We need to look at how we can work with industry to manage issues together, rather than us saying 'this is how it must be, you must do this, you must do that'. One of my aims is to explore how can we get shared responsibility, shared ownership, improved communications" (Oliver 2015a).

In our fieldwork, we found that fishers perceive the greatest threat to the sustainability of the inshore to be inadequate participatory mechanisms hindering their ability to engage in, and contribute in a meaningful manner to, the managerial process. They are particularly concerned with the current lack of feedback from authorities on suggestions they make to improve inshore fisheries management. This was at first sight a surprising finding because lack of access to quota appears to be the main threat, since this is the most frequently discussed topic in the sector-specific press, and the major campaigns mounted by fishers centred on the inability of fishers to make a living through quota allocations. What our research revealed, however, is that inadequate quota is a symptom of an ineffective participatory infrastructure, rather than the primary cause of discontent. The main way that English SSF perceive themselves to be vulnerable is because they feel powerless to deal with the threats facing them as they are not listened to and marginalised since inappropriate management pathways thwart their meaningful participation in fisheries management decision-making. Many fishers refer to barriers preventing their access to decision-making bodies such as IFCA committees, citing the complex and overly bureaucratic application process which they struggle with. Vacancies on IFCA committees are claimed to be poorly advertised, and fishers criticise the way they perceive the MMO controls the application process and discriminates against certain individuals and communities. Moreover, the timing and location of IFCA committee meetings, given fishers' work-loads, serve as further obstacles, since fishers say they would lose a day's work with inadequate compensation. Also the format of meetings discriminates against fishers, demanding a degree of professionalism which is alien to the everyday experiences of most inshore fishers. Meetings are often seen as tick-box exercises, offering little opportunity for genuine debate, and fishers say they struggle with the heavy paperwork requirements of IFCA committee membership. Likewise, fishers are suspicious of the one-on-one meetings offered to them by both the MMO and the IFCA, believing those meetings are allocated disproportionately to fishers favoured by the organisations. Meetings targeting 'key stakeholders' in a community are also criticised because they usually involve the head of the local fishermen's associations, which is no guarantee of reaching the wider community since many associations are fractured. Several fishers see the shift by their IFCAs away from physical meetings to a reliance on uploading information to their website as problematic and in some cases discriminatory, because fishers who do not own a computer or are uncomfortable with computers are disadvantaged.

The consequences of this perceived lack of participation in fisheries management decision bodies are far-reaching. For example, barriers to access mean IFCAs can struggle to recruit fishers to their committees. As a result, groups such as the environmental NGOs, marine developers and producer organisations (POs) which can put forward paid and experienced members gain an advantage to lobby and influence the system. This imbalance is exacerbated by the presence of local authority councillors on IFCA committees, immediately assigned seats given their council status to oversee their councils' financial investment: on average they take up 47% of the voting space. Their inexperience, limited enthusiasm and lack of sector-specific knowledge makes councillors easily manipulated by dominant officials. The result is a poor fit between management regulations and the needs of the industry. The creation of regulations without relevant grass-roots input means that some rules are inappropriate for the site for which they are designed: for example, they are unsympathetic to the traditional seasonality of local fishing. Fishers report that management risks being divorced from the reality on the ground both in terms of the practicalities of fishing and the site- specificity of the fisheries. For example, quota often became available to fishers at a time the fish are absent from their areas, and quota are often withdrawn at very short notice.

Another consequence of lack of participation is that fishers' knowledge is not made use of in management decisions. Managers do not routinely draw on fisher's knowledge of their grounds and target fisheries, but instead rely on data their scientists collect or commission. Fishers' knowledge is derived from the practical experience of fishing activities collected over many years, some of which may be recorded in log books and other documents, but much more of it is stored in fishers' memories. It is well-established that people whose lives are associated with living resources and marine ecosystems acquire substantial knowledge about ecosystem relationships, the status of species in the ecosystem and the interactions with human activities such as fishing. Fishers are very aware, for example, of natural fluctuations in the abundance and distribution of fish, although their knowledge is local and personal or subjective. Fishers claim that for some species and areas, the information available to them is better than the information available to scientists, and they distrust the simple models employed by scientists, and question the validity of the data collected. For their part, some managers distrust fisher knowledge because fishers are not trained scientists. This mutual distrust arises in part from the different perspectives of inshore fishers and scientists, and there is more often than not no arena available for them to discuss these differences openly and honestly: at present, inshore fishers have very limited roles in the English science review and advisory processes. This failure erodes fishers' trust in both the system and the resulting data. The lack of any means for incorporating inshore fishers' knowledge and experience into the traditional scientific advice is a major weakness of the current system of stock assessment in the English inshore sector, impoverishing the data from which decisions are made. Evans and Andrew (2011) make the point that in developing countries, where data on SSF are very poor, it is even more important to have fishers' participation in management decisions.

Lack of participation also leads to an increase in non-compliance, an outcome that is in line with other studies which testify that without participatory opportunities the chances of fisheries regulations succeeding are poor as fishers will breach regulations. This happens in two ways. First, a lack of participation opportunities means that fishers do not always understand why regulations have been imposed, and there is little sense of ownership at a grass-roots level. Second, fishers report that they are subject to many overlapping regulations which, viewed in combination, become unsuitable and unworkable, compliance with which would make them bankrupt. An array of management measures has been imposed upon the U-10 m fleet in England since 2005, chief amongst them being controls on fishing outputs such as catch limitations which, when operating in a mixed fishery, cause high levels of discards. Bureaucratic tendencies to think compartmentally have led to overlapping regulations created in isolation, as seen with the recent discard ban regulations implemented with quota systems still in place, which makes the landing obligation practically impossible for SSF to comply with. The discard ban or landing obligation introduced by the EU in 2014 is directly attributed to the campaign by Fearnley-Whittingstall, and is expected to harm the U-10 m sector much more than the O-10 m sector because smaller vessels lack the storage space for large quantities of unwanted fish that have to be transported to shore, and they cannot so easily avoid catching 'choke' species which could end a fishing trip (Veiga et al. 2016). Fishers say they are forced to break the rules if that is necessary to keep their business alive, but they do so with a heavy heart coupled with depression and anger as they feel pushed into criminality. Jerry Percy, director of New Under Ten Fishermen's Association (NUTFA), said inshore fishers are faced with the choice of either "criminality or bankruptcy" unless the authorities change the landing obligation rules for SSF (Oliver 2019g).

Lack of trust in management also undermines fishers' faith in the integrity of the decision-making process and in the management groups involved in this process. Some fishers express an ingrained suspicion that rules are constructed behind closed doors under the influence of groups with an anti-fishing agenda. Some local fleets do not wish to engage with a system which they perceive as hostile. This is in line with the literature which finds that fishers will only take advantage of participatory mechanisms, however easy they are to join, if they feel that in doing so they will be influential (Pelling 1998). This study also finds that lack of stakeholder participation adversely affects fisher well-being, including deterioration in their confidence and self-worth (Béné 2003).

On the other hand, our research found that some respondents praise their IFCAs for listening to them and consulting them before making decisions, while some fishers report helpful responses from the MMO. These positive assessments of IFCAs and the MMO in relation to stakeholder participation are replicated in two reviews of these organizations produced by Defra in 2015 and 2014 respectively. In its 2015 report on IFCAs, Defra concluded that most IFCAs fulfilled Defra guidance that "IFCAs work in partnership and are engaged with their stakeholders" (Defra 2015, p. 17). In its 2014 report on the MMO, Defra said that customer satisfaction with its services was 70% in 2013, and 65% in 2014 (Defra 2014). The editor of *Fishing News,* Dave Linkie, sums up these opposing perceptions of the work of the IFCAs and the MMO as follows:

> "The industry's perception is that the IFCAs are loaded against them and are institutionally biased in favour of achieving environmental objectives rather than looking after the interests of fishermen. The MMO/IFCA perception is that the system is set up to ensure that fishermen's views are fully represented and that the appointments procedures tick all the boxes to ensure that this is so" (Linkie 2016).

However, SSF participation in fisheries management is not without its critics. Some critics focus on the practicalities of implementing stakeholder participation, arguing that it is very difficult to decide which stakeholders should participate, and what powers they should be granted. A common prescription put forward to give effect to SP is co-management, and there is an increasing number of co-management arrangements for SSF across the world. However, Gelcich et al. (2006) claim that co-management arrangements reduced resilience in traditionally managed fisheries in Chile. Davis and Ruddle (2012) argue that some forms of co-management co-opt SSF into the prevailing neo-liberal agenda of modern industrial fisheries management, at the cost of cultural values and norms of traditional fisheries. Finkbeiner and Basurto (2015, p. 434) say that "Similar to other governance configurations, co-management can result in unintended consequences—often the usurpation of political power by private or special interests". According to Béné (2009b, p. 253), the real beneficiaries of co-management are rarely the SSF but usually the most powerful local elites. He claims that his review of the implementation of co-management programmes in sub-Saharan Africa during the previous 20 years showed that "in most cases, fisheries co-management programmes failed to improve governance, but simply altered the distribution of power and responsibility amongst the different stakeholders". Béné (2009b) claims that critics of co-management blame this failure on an inadequate amount of power devolved to SSF: in other words, there was an insufficient degree of stakeholder participation, probably because managers were reluctant to devolve power to SSF. However, Béné (2009b, p. 259) himself argues that the problem is not so much the inadequate amount of participation as the type and basis of participation: "Thus, advocating for a systematic strong participation by the fishery community may not be the correct approach and the issue of *how much* power is shared may be the wrong question. Instead, issues of *how* this power is shared and *who* receive(s) this power may be more important" [Italics in original]. Research he carried out into the relation between the amount of participation and the success of SSF found that "the degree of participation did not explain the performance of the fisheries: some fisheries characterized by highly centralized management systems were doing well, while other, more participatory, fisheries were unable to generate good management outcomes…'More participation' is not the panacea" (Béné 2009b, pp. 258, 263). Accordingly, Béné et al. (2007) suggest several conditions that must be met before co-management would fulfil the requirements of stakeholder participation for SSF. There must be 'downward accountability' of representatives to their communities, legislative backing, political support, local and central governmental collaboration, rules to prevent elite capture, adequate funding, and clarity of objectives. We address some of these conditions in the next section in our recommendations for the governance of the English SSF.

7.4 Recommendations for the English Inshore Sector

Our first recommendation for improving the challenging situation of the inshore fleet is for management to introduce more effective participatory mechanisms for managing these communities' working operations. This is a demanding requirement with no guarantee of success, but if management recognise the long-term benefits they themselves will reap from better relations with fishers, it has a chance of working. IFCAs have more conservation responsibilities than did their predecessors, the Sea Fisheries Committees (SFCs), and they have lost much of their autonomy since other governing bodies, including the MMO, now influence their work-streams, and they are working with reduced budgets to implement wider portfolios. It is tempting, therefore, for IFCAs to downgrade consultation exercises, but excluding fishers is likely to increase their workload in the long-term in enforcing conservation regulations on a disgruntled workforce. This study argues that it is in management's own interest to dedicate resources to understand how different communities work and develop proactive engagement strategies to work along the grain of their communities' aspirations rather than against them. As part of these proactive consultation processes, managers need to explain why some measures are going through and other measures are not, in an open, honest and frank manner. Where this has occurred, there has been a positive response from fishers which has made the IFCA's job easier, demonstrating that their resource-intense consultation exercises have paid off.

In our view, a central feature of an IFCA's responsibility is to be on high alert to detect signs of communities starting to move towards negative passive resilience strategies. This study has provided management with key indicators which can be used to predict when communities are starting to move in unsustainable directions. For example, when fishers stop attending meetings, alarm bells should be ringing and managers need to take steps to identify the cause and work out how to encourage fishers to come back on board. Furthermore, IFCAs need to know how to nurture transformative and adaptive initiatives, so when they come across a source of optimism and a desire to cooperate, they must take advantage of it, or if they are unable to do so at the time, they must explain why. Key indicators of fishers' disillusion would be picked up by managers much earlier if fishers were included in such decision-making processes.

This recommendation for more effective participatory mechanisms is not a radical or revolutionary demand, but merely a request that IFCAs fulfil one of their already-agreed responsibilities (Cappell n.d.). Our recommendation is also in line with the vision on the future direction of UK fisheries expressed in Defra's 2018 White Paper on *Sustainable fisheries for future generations*, which links more stakeholder participation with an increased contribution from the industry to help defray the cost of fisheries management: "Our future vision is that industry should take a greater, shared responsibility for sustainably managing fisheries, while making a greater contribution towards the costs. This can, for example, work to develop new management practices and contributing to fisheries science, being part of the delegation in the negotiations, being more actively engaged in fisheries management decisions and co-designing future policy" (Defra 2018, p. 38). It is also in line with the broader vision of the reformed EU CFP, as expressed by Fisheries Commissioner Maria Damanaki in 2013 (Arias-Schreiber et al. 2019). We are not suggesting that management roll over and accept all the suggestions fishers make to them, but that in committees (including IFCA committees) and other forums, fishers and managers meet on much more equal terms so genuine and open debate takes place between them. In these meetings, IFCAs and the MMO need to demonstrate that they are listening and providing full and evidence-based reasons for not taking up suggestions.

In order to make such participation viable, however, government must take steps to help build up the capacity of fishers and their organisations and communities to engage effectively in management decision-making. This entails not only education and training of fishers to present their views more authoritatively at management meetings, but also efforts to reinforce local social capital by reducing "disadvantageous power differentials that exist in many fishing communities in developing countries" (Kosamu 2015, p. 372). SSF participants must be chosen by fishing groups and communities not by management; arrangements must be made for them to attend decision-making meetings at times and in places convenient for them; and training should be given to fisher representatives to equip them to play meaningful roles in discussions. Agendas for fisheries management meetings must not be manipulated by management or elites to exclude SSF concerns from consideration (SSF representatives should have the right to add items), nor should discussions be monopolised by dominant individuals or groups. Management must be held to account for their actions by explaining why decisions were made (or not made) and what the consequences were, and SSF participants must be treated with respect and dignity, and genuinely listened to by management (FAO 2015).

There is also a gender dimension to this participatory recommendation: women should be encouraged to play a bigger role in inshore fisheries decision-making processes. We regret that our study did not pay sufficient attention to the roles played by women in the English SSF in helping fishers to carry out adaptive or transformative strategies. We were made aware of the jobs done by some women working behind the scenes and of the activities of a few women in leadership positions (e.g., FA leads). But we failed to ask questions about the significance of these female roles in contributing to the resilience of the SSF. However, evidence from other studies suggests that women can make a significant difference to the efficiency of inshore fisheries operations and to the effectiveness of inshore fisheries management (Fröcklin et al. 2013; Sharma 2004). Therefore, it is key that fisher-women's roles need to be recognised, understood and valued. Women contribute significantly to the English SSF but are under-represented with little say in decision-making about SSF at all levels. Given their important place within the SSF, we need to include women in management and ensure their equitable access to financial support and appropriate training on income-generating and management activities (Britton 2012b).

Our second recommendation follows on from the participatory requirement of our first recommendation - to encourage partnerships between SSF, scientists and managers to conduct collaborative research. During the last 20 years, considerable collaborative research has been carried out between large-scale fishers and scientists—fisheries/science partnerships (Mangi et al. 2018). But there has been very little collaborative research carried out by small-scale fishers and scientists. Trimble (2014, p. 98) describes such participatory research as

> "a knowledge co-production approach with an action-oriented component based on local interests and concerns, in which local people participate in the entire research process, and whose final aim is community empowerment…Participatory research has become increasingly common in the context of natural resources and environmental management…including fisheries…It offers one way to create power-sharing relationships between researchers and communities, to develop locally appropriate resource management strategies, and to strengthen social relationships".

In our view, the lack of collaborative research between SSF and scientists is a missed opportunity for five reasons. First, SSF could conduct very valuable mapping work of marine resources especially in the benthos: at present, there are huge gaps in our knowledge of the coastal benthic system. Second, it could improve relations between SSF and scientists as well as managers who would oversee and fund the joint research. Working together invariably generates mutual respect (Mackinson and Wilson 2014). Third, it would enable SSF to develop a deeper understanding of the fragility of the marine ecosystem, and encourage them to become environmental stewards of the seas. Fourth, it would provide extra income for SSF. Fifth, it would increase community empowerment (Trimble 2014). Pena and McConney et al. (2014) add another task for such joint research: in addition to obtaining data on the marine ecosystem, fishers could be involved in socio-economic monitoring.

Our third recommendation is for management to be more flexible and less rigid in their decision-making. Adaptive management is the name of a smarter way to manage SSF. As Bown et al. (2013) note, adaptive management acknowledges that socio-ecological systems are too complex to be fully understood and therefore cannot be managed by conventional methods which assume they are intelligible and predictable. Instead, adaptive management accepts the need to adapt to perpetual change rather than to strive for a mythical constancy or stability. This means managers must be nimble and light on their feet in responding to changed circumstances affecting the SSF. They should be willing to experiment with new and innovative ideas, and be prepared to learn by doing, thinking outside the box, rather than insist on enforcing traditional rules and regulations which have clearly failed. Jentoft et al. (2011, p. 464) call for more flexibility from managers: "governance needs to follow what we call the *dexterity principle*, by which we mean sensitivity to details that differ from one situation to another. It also means taking into account how people are actually living and operating, and what they value" [italics in original]. McClanahan et al. (2009) recommend trial and error. McConney et al. (2014, pp. 3, 5) say "SSF are complex adaptive social-ecological systems. Far from trying to control such systems, the approach to stewardship must rely on our ability to develop strategies for understanding and adapting to the complex, unpredictable and emergent properties of such systems...Fisheries and environmental policies centred on trying to control systems have led to increased unsustainability and social crisis". As Parks (2011, p. 95) puts it, management decisions should be "based on observation and experimental evidence rather than a priori reasoning". Ratner and Allison (2012, p. 389) argue that "proponents of reform should avoid overconfidence in the effectiveness of governing institutions...and place a premium on learning through institutional experimentation". Aguilera et al. (2015) assert that management flexibility is the key to facilitating SSF adaptability. One way of making IFCAs more flexible is to reduce the degree of control exercised over them by the MMO, and restore the autonomy that their predecessors—the SFCs—enjoyed.

Our fourth recommendation is to fishers: inshore fishers and their communities should respond positively to overtures from the IFCAs. They should not squander management goodwill by becoming divided among themselves. In this research, we have found the inshore fleet as a whole to be more resilient than we expected, and we are guardedly optimistic about its future. But a major element of its sustainability is its determination to work with, rather against, management. The well-being of the English SSF, therefore, depends on both the political will of management and the reciprocal spirit of cooperation in fishing communities. Moreover, these two factors are connected: as Carbonetti et al. (2014) point out, the political will of authorities can be reinforced by local champions of SSF.

Our fifth recommendation is for government to address the current maldistribution of quota between the industrial sector and the under-10 m sector. They should re-apportion quota between the offshore and inshore sectors (excluding the super U-10s), to give a fairer share to the U-10 m fleet. The FAO Guidelines "support equitable distribution of the benefits yielded from responsible management of fisheries" (FAO 2015, p. 5). Defenders of the quota system argue that quotas are now traded in markets as commodities, so quota owners regard them as property rights. However, as Robert Nozick observed in his book *Anarchy, State, and Utopia* (1974) on retributive justice, if a commodity was originally acquired unjustly, no matter how many times it is subsequently traded, it remains an unjust holding. Defra's 2018 fisheries White Paper does not, however, give much confidence that the government is likely to carry out such a redistributive policy: "We do not intend to change the method for allocating existing quota...Fixed quota allocations (FQAs) were established in 1999, based on a reference period of 1993–1996, and we recognise that fishermen have invested in FQAs. We will continue to use this methodology for the apportionment of existing quota" (Defra 2018, pp. 12; 26). Huggins (2018) comments "there is a sense that the concerns of the small-scale fishing fleet have been ignored". An amendment to the Fisheries Bill (setting out post-Brexit fisheries arrangements) currently (January 2020) going through the House of Commons, put forward by several MPs including two former fisheries ministers, sought to change the criterion for allocating quota from the historical catch records of the existing FQA system to criteria based on the environmental impact of fishing, and fishing's social and economic contribution to local communities (Oliver 2019a). NFFO (2019a) campaigned against the amendment on grounds that the 20-year old FQA system had brought stability to the industry and radically improved the health of most fish stocks in UK waters. NFFO (2019a) claimed the problems of lack of quota for SSF were not due to the FQA system but to the fact that the super under-10 m vessels which constitute 14% of the total UK SSF fleet take 70% of the quota allocated to the UK under-10 m pool. Also, NFFO proposed that the U-10 m fleet should be cut by 50% (just like the O-10 m fleet had been cut during the last 20 years) to give the remaining vessels sufficient quota to survive. What NFFO fails to acknowledge is that decommissioning half of the U-10 m fleet could undermine the viability of SSF coastal communities. The SSF can draw some comfort from the government's stated intention of discussing with stakeholders how to allocate future additional quota that might become available as a result of Brexit from December 2020 onwards, using the geographical criterion of zonal attachment rather than the historical criterion of catch records during reference periods (Defra 2018). In a response to this consultation initiative, NFFO (2019b, pp. 3, 4) expresses some sympathy for two scenarios favourable to the SSF: first, "The allocation of additional quota could be made contingent on bringing long-term benefits to a specific region/coastal community"; and second, "There is a strong argument for taking genuine low impact vessels out of the quota management system altogether by accounting for the mortality attributable to their catches by a single allocation at the start of each year". These are encouraging sentiments, but NFFO warns against fuelling an undue expansion of the inshore fleet because it is already an overcrowded sector and has a "significant latent capacity issue" (NFFO, 2019b, p. 4).

On the other hand, we recognise the point made by Song et al. (2018, p. 291) that a redistribution of quota from industrial fishers to SSFs would only address the symptoms of the problem and leave the source of the problem—the whole neoliberal system of FQAs (in effect, ITQs)—fully intact: indeed, it reinforces that system by giving it greater apparent legitimacy, thereby undermining the prospects of its abolition and replacement by effort control and/or place-based regulations.

Our sixth recommendation is for government to introduce exclusive fishing zones (EFZs) for SSF using non-destructive gear in England. Béné et al. (2015a, p. 41) say that "A positive response to the threat of competition for small-scale fishers is to introduce preferential fishing zones from which commercial operators are excluded". As Ramírez-Luna and Chuenpagdee (2019, pp. 198, 213) note, "placed-based management tools such as EFZs can be a powerful mechanism for the protection of the rights of small-scale fisheries and fish workers around the world". Leis et al. (2019) report the success of EFZs for SSF in Southern Brazil. De Mattos and Wojciechowski (2019, p. 502) refer to this development as "re-artisanalisation" of fisheries activities. Courtney et al. (2016, p. vi) describe a similar situation in the Philippines, whereby exclusive access to inshore waters extending to 15 km of the shoreline has been enshrined in national law. Our proposal is to restrict commercial fishing within six nautical miles from the coast of England exclusively to the English SSF (not including the super-under 10 m vessels). At present, in certain circumstances, industrial fishing vessels (as well as super-under 10s) are allowed access to these inshore waters, and we believe this should cease in order to protect the 'genuine' SSF fleet from potential extinction.

Our seventh recommendation is to both government and fishers—to promote Marine Stewardship Council (MSC) accreditation for SSF. MSC accreditation will add value to SSF catches, but pre-assessment it is too costly for most SSFs, so government assistance is needed (Davies et al. 2018). Duggan and Kochen (2016) explain that MSC is often accused of being out of reach of SSF: only 2% of MSC accreditation is for SSF. Project Inshore, an initiative led by Seafish during 2012–2015, mapped more than 450 fisheries in the English inshore sector, and assessed them against MSC criteria of sustainability (Nimmo and Southall 2012). The project found that about 50 fisheries were performing at a level that suggested they could be put forward as candidates for full MSC assessment (Hjul 2015). We recommend that public funds be provided to pay for full MSC accreditation for these 50 fisheries. On the other hand, we are aware of the argument put forward by Penca (2019) that the focus of MSC accreditation is purely environmental and does not take into account social and community considerations of SSF. Instead, Penca (2019) argues for locally-based traceability schemes customised for SSF's socio-economic circumstances.

Behind many of these recommendations lies our opinion that the English SSF deserves support from the British government in recognition for the 'public goods' it provides. The concept of 'public goods' refers to benefits that are provided by producers free to the rest of society (Ferdman 2017; Holcombe 2000). The implication is that anyone who provides such public goods deserves to be subsidised by the state: "public goods need to be sustained—at least to some degree—by means of public finance" (Kallhoff 2014, p. 642). The British Government already recognises an obligation to recompense farmers for the public goods they provide. Michael Gove, when Environmental Secretary, listed 12 public goods provided by farmers, including improved water, air, and soil quality, increased biodiversity, enhanced beauty of the natural environment, world-class animal welfare and health standards, preservation of rural resilience and traditional farming and landscapes, and facilitation of public access to the countryside (Defra 2018). The Agriculture Bill currently going through the British House of Commons sets out how farmers and land managers will in future be paid for such 'public goods'. We suggest that a comparable list of public goods provided by the English SSF can be drawn up, including increasing our knowledge of fish stocks and marine fauna and flora; monitoring the health of marine ecosystems; reporting violations of marine regulations; rescuing distressed sea-goers; manning lifeboats; supporting coastal communities' struggling economies, and upholding local traditions, heritage and tourism—all of which justify governmental support. We recognise that the English SSF already receives some governmental support (though not as much as does the large-scale sector), including loans and grants for new and refurbished boats and equipment; tolerance of some negative environmental impacts; and some allocations of fish quota beyond scientists' recommended levels. We also recognise that the English SSF (like small farmers) produces some public 'bads' (though not as many as the large-scale fisheries sector does (NUTFA 2014; Oliver 2019f)), such as benthic damage, overfishing, ghost nets, marine litter and CO^2 emissions. But we contend that the public goods provided by the English SSF far outweigh these public bads.

7.5 Wider Implications

This examination of small-scale fisher (SSF)'s resilience strategies across England has relevance for the global small-scale fisheries problem. SSFs around the world are marginalised and vulnerable, confronted with a broad range of threats including natural hazards, marine developments displacing them from their traditional fishing grounds, competition from large numbers of illegal, unreported and unregulated (IUU) vessels, and neo-liberal policies of individual transferable quotas (ITQs). This last threat - "a worldwide trend to gradually introduce rights-based fisheries management systems for regulating access to marine resources" (Sowman 2006, p. 70)—is one of the most worrying problems for SSF, since it threatens to price SSF permanently out of fishing opportunities. Song et al. (2018) say the ITQ system has now established an ideological grip on contemporary fisheries management, and they analogise this privatisation of the marine commons to the enclosure of the terrestrial commons in England in the late Middle Ages. Crilly and Esteban (2013, p. 26) state that "It remains a possibility that in a fully privatised fishery, the stock could be driven to extinction for profit". Faced by such an existential threat from industrial fisheries, many SSF seem powerless to protect themselves, not least because of their lack of participation in fisheries management decision-making. What they need is support from outside their own ranks to obtain a greater voice in fisheries management in order to secure more equitable access to fisheries resources. Such external support has begun to materialise during the last 15 years partly as a result of growing academic interest in the SSF's predicament.

SSFs are increasingly researched, as indicated in the burgeoning literature on the subject, encouraged by the establishment of the peer-reviewed Centre for Maritime Research (MARE) Publication Series, hosted by Springer, which strives to elevate the profile of small-scale fisheries around the world. SSFs have attracted considerable international interest and support during the last few years since the benefits of coastal fishing communities have become better understood as food and livelihoods providers and as contributors to cultural heritage, social cohesion, and identity. This momentum resulted in the creation of the research conglomerate, Too Big To Ignore (TBTI), in 2012 and the FAO's Committee of Fisheries endorsement of the Voluntary Guidelines for Securing Sustainable Small-Scale Fisheries in the context of Food Security and Poverty Eradication (SSF Guidelines) in 2014. These Guidelines mark a historical moment for millions of small-scale fishing people around the world: never before has this sector received such global recognition. A number of countries have already taken steps to implement the SSF Guidelines, including Algeria, which launched a 'Chartre' for sustainable fisheries and aquaculture which specifically refers to the SSF Guidelines in 2014. Costa Rica enacted an executive decree in 2015 regarding the official application of the SSF Guidelines. Malta established a permanent working group within the General Fisheries Commission for the Mediterranean in 2016 with the prime objective of facilitating the implementation of the SSF Guidelines. The Pan-European grass-roots movement, Low Impact Fishers of Europe (LIFE), which was formed in 2012 to represent SSF at EU forums and to coordinate efforts to promote the interests of SSF in countries across Europe, has attracted an impressive number of members and issues a monthly newsletter reporting progress of their initiatives which defend SSF (LIFE 2016). Nayak and Berkes (2019, 206) say that "the relatively recent emphasis on small-scale fisheries (as opposed to large-scale fisheries, usually privileged by governments) provides a window of opportunity to rebalance fisheries policies at the national and other levels".

SSFs across the world employ a wide variety of coping mechanisms, but these often fail for similar reasons to some fishers in the English fleet, including lack of the leadership capabilities needed for bringing them out of their challenging situation. Amongst other things, the SSF Guidelines seek to address this key point, and also emphasise how effective participation by fishers in designing effective and fair management measures is required, since whilst the SSFs need a voice, this is not enough: they must be provided with actual control over the conditions under which they work. The Guidelines stress that provision must be made to build capacity in those communities not able to engage, and methodologies need to be provided for association building, developing trust, mutual understanding, and institutional arrangements that enable respectful multi-directional communication and exchange (Charles 2011a). The present book strongly endorses this call, stressing that achieving fisher participation in management is not as simple as providing the opportunity and waiting for fishers to turn up. SSF fishing communities are enormously diverse, and management must understand that fishers cannot be regarded as a single group: one size does not fit all, and each community must be treated as sui generis (Symes et al. 2015; Jentoft and Chuenpagdee 2015a). Authorities must get to know their communities in order to tailor their engagement processes to the mind-sets of the fishers they are dealing with (Jentoft and Chuenpagdee 2015b; Jentoft and Eide 2011).

The fact is, as Jentoft et al. (2011, p. 452) put it, "Small scale fisheries are too big to ignore…Small-scale fisheries must be elevated on the political agenda". Carvalho et al. (2011) urge governments to give SSF 'top priority'. Sowman and Cardoso (2010, p. 1169) call for a renegotiation of the "conflicting interests of powerful well-resourced (industrial and recreational) groupings and the impoverished poor under-resourced small-scale fisheries sector". We endorse the sentiment of Jentoft et al. (2011, p. 467) that given the right circumstances, "Small-scale fishing can…provide a good life". As Pauly (2006, p. 7) said, "despite their present problems, the small-scale fisheries of the world, suitably governed, are still our best hope for sustainable utilisation of coastal resources". Indeed, Pauly (2018, p. 371, 373) claims that the 'Blue Economy' depends much more on the SSF than on industrial fishing: "While industrial fisheries tend to lack the features that would make them compatible with a blue economy, small-scale fisheries possess most of these features, and thus may represent the future of sustainable fisheries…Indeed, if fisheries have a long-term future, it will be as local activities, embedded in a blue economy, the complement of a green economy on land". So, far from large-scale fisheries replacing small-scale fisheries, Pauly's vision (p. 373) of the future is SSF replacing industrial fisheries: "in many countries, well-monitored artisanal fisheries could, at least partly replace the industrial fisheries with which they currently compete". As Stobberup et al. (2017, p. 15) note, however, the EU has omitted fisheries (including SSF) from its 'Blue Growth Strategy', "presumably because it is perceived as having limited potential for growth". The fact is that the notion of the Blue Economy or Blue Growth is the latest contested arena in which SSF must make its case heard. As Cohen et al. (2019, pp. 2, 4, 6) point out,

> "There are concerns expressed by small-scale fisher groups that the Blue Economy agenda undervalues social objectives, and in doing so threatens the basic imperative of providing both livelihoods and affordable, nutrient dense food for those who need it most…To date, considerations of food security and human rights have not been front and center in high-level dialogue around the Blue Economy. Small-scale fishers have been notably underrepresented…If the Blue Economy is to be a legitimate vision for governing the oceans, then alongside industry and conservationists, the voices, interests and human rights of the largest groups of ocean-users – women and men who service, fish and trade from SSF – must be represented and recognised from the outset of the solution design".

For Percy (2016), SSF is part of the solution, not part of the problem: "smaller scale fishermen and women are not a problem to be ignored but with recognition, support and fair access to resources, can be a solution to some of the many problems facing the marine environment today". In the case of the English SSF, Seafish's 2012-2015 survey of 450 English inshore fisheries using the MSC's sustainability criteria to assess their stock, environmental and management statuses, provided bespoke 'roadmaps' for sustainability for each of the 10 IFCAs (*Fishnews* 2015). The advisory group for this initiative, dubbed 'Project Inshore', was chaired by MSC's Claire Pescod, who claimed its interim report "highlights best practice in England's inshore fleet but also illustrates a huge need for investment in our inshore fisheries, a fleet that has provided fish for generations. By investing in the science behind fisheries, we are working towards not only sustainable fisheries but also sustainable livelihoods and communities for generations to come" (*Fish Site* 2013). Seafish has followed this initiative up with the formation of a steering group known as the Future of our Inshore Fisheries, which organised a major two-day conference on the future of UK inshore fisheries in October 2019. According to the NFFO, the core question for this conference was "what framework will allow the fishing industry to take more responsibility for managing its own fisheries, sector by sector, area by area" (Oliver 2019b). Michel Kaiser, who chaired the conference, said a central point that emerged from it was the necessity of co-management between the under-10m fleet and the government—i.e., for the inshore sector to be "legitimately part of the decision-making process" (Kaiser 2019). Linkie (2019a) states "There was a general consensus that the inshore sector must be centrally involved in the management of its own industry, which will require changes in the current institutional management arrangements". Oliver (2019c) notes that this conference comes at a time when "the new fisheries bill will give the government powers to set up a completely new fisheries regime once the UK leaves the EU and the CFP", and he asserts that the conference "was the first step towards a new era of co-management of inshore fisheries between fishermen, officials, and scientists" (Oliver 2019d). In addition, in December 2019, a cross-industry working group was set up at a meeting in London to raise the profile of the shellfish sector in England, Wales, Northern Ireland and the Channel Islands, and "The meeting agreed that the small-scale sector of the shellfish fleet sometimes finds it difficult to make its voice heard, and that this must be addressed" (Oliver 2019e). The British All-Party Parliamentary Group on fisheries said "The momentum surrounding the conference…must be put to good use" (Oliver 2019f). Linkie makes the interesting point that these moves towards co-management of the inshore sector have been stimulated by Brexit: "Brexit has obviously been a catalyst for these developments, and will give the UK a unique opportunity to install a new fisheries management regime using the powers in the new fisheries bill" (Linkie 2019c). So, although Brexit may not give the under-10m fleet more quota, it may lead to more participation for inshore fishers in SSF management.

Support for SSF is, therefore, not looking backward to a mythical past, but forward to a rational future. As Cohen et al. (2019, p. 5) point out,

> "Resilient SSF have adapted and modernised, and in many instances are both sophisticated and highly efficient…Despite some SSF having long histories and cultural connections, SSF are not necessarily antiquated or outmoded, and cannot be dismissed simplistically as historical relicts of a bygone age. Small-scale fishers in poor countries have been early adopters of technologies…and have responded to demands from new markets…The dynamic nature of SSF has seen them persist despite ever-increasing and diverse pressures".

But it is a future that is inextricably bound up with politics. Referring to Article 17 of the reformed CFP which requires Member States to take into account socio-economic and environmental factors when allocating access to fish quota, Percy (2016) asserts that it all boils down to a question of political and public will: "If governments took this approach seriously, rather than only pay lip service to it, if at all, then it would revolutionise the industry very quickly…but until this happens then the system will continue to reward those who fish the most rather than those who fish the most sustainably". Three years later, Percy says the EU inshore fleets are still waiting for the political will of member states to materialise: "The SSCF is the 'forgotten fleet' because no member state has embraced the requirements of the Article 17 of the CFP…and few, if any, have developed the required action plan for the small-scale fleet" (Oliver 2019g).

As Campling et al. (2012, p. 192) reminds us, fisheries management is a "political process"; issues are fought out on a "site of political contestation"; and an argument for SSF access to fishery resources is a "political endeavour". There is nothing inevitable about the dominance of industrial fisheries: it is the result of a conscious political choice to adopt a neoliberal rather than a communitarian agenda. As Jentoft remarks, "States must take the lead and enable the implementation of the SSF Guidelines, even if it clashes with their dominant political paradigm. The SSF Guidelines advocate for a massive paradigm shift, and implementation will likely not be an easy process" (TBTI 2017, p. 10). Thornton and Hebert (2015, p. 370) explain how neoliberal fisheries managers seek to disguise the political nature of their policies by dressing them up as technical matters: "the 'antipolitics' informing modern bureaucratic regimes continues to frame management problems around 'closed', seemingly depoliticised technical issues" such as MSY. But this is a smokescreen to divert attention from the political basis of such technical discourses. Andrew and Evans (2011, p. 19) claim that "Property right issues still dominate fisheries management"; Tolley and Hall-Arber (2015, p. 406) say "Neoliberal economic approaches to fisheries policy...currently dominate in policy and public discourses and action"; and Pinkerton (2017, p. 6) warns that "neoliberal policies in fisheries are becoming more widespread". However, Pinkerton (2017, pp. 6, 3) also says that there are "instances of its being in retreat or being bypassed by alternatives", and she reminds us that "It is quite possible for the state to limit neoliberalism, often by simple regulations, in ways which protect the welfare of SSFs". For example, the state could guarantee a fair allocation of quota to SSF and prevent it being sold to industrial conglomerates; it could forbid trade in SSF fishing licences by treating them as public property to be surrendered to the government when a fisher ceases to fish; and it could exclude industrial vessels from fishing close to shore. Moreover, Pinkerton (2017, p. 8) describes the efforts of social movements led by NGOs and fishers' associations to fight for SSF rights and concludes that "neoliberalism is definitely encountering greater challenges...Some SSFs are becoming better organized and better able to protect their interests...Despite the deep claws of neoliberalism embedded in the thinking and practices in fisheries management worldwide...SSFs are becoming better understood and respected for their contribution to public welfare, and...they are likely to retain and expand their place". Pinkerton and Davis (2015, p. 308) point out that "the implementation of neoliberalism agendas in fisheries is not a fait accompli, but an ever unfolding struggle", and there is "a diverse group of actors...which have...managed to disrupt the efforts of neoliberal planners". As Pitchon (1997) puts it, "Politicians made these rules, *and they can change them.* There is nothing inevitable about 'unfettered' free markets...The logic of corporations cannot become the logic of nations. Markets have the power that they have, and corporations who benefit from these markets have the power that they have, because politicians everywhere have acquiesced in this project" [Italics in original].

Sowman et al. (2014, p. 31) report how in South Africa, a seven-year campaign driven by SSF, civil society, NGOs and researchers resulted in "a paradigm shift in the governance of small-scale fisheries...from a largely resource-centred approach to one that is more people-centred and which recognises fisher rights as human rights". In Japan, SSF is thriving because the government reversed its previous flirtation with Western neo-liberal privatisation measures and returned to the traditional community sea tenure system, empowering SSF in challenging large-scale operators, thus demonstrating that path dependencies are reversible (Jentoft and Chuenpagdee 2017). Witter and Stoll (2017, p. 132) record how several measures have been taken to safeguard SSFs in North America: "policy mechanisms have been implemented to help enhance access opportunities for small-scale fishers operating within quota systems…In the US, the Magnuson-Stevens Act (2007) introduced Community Fishery Associations as entities that communities can use to hold fishing quota". Stewart and Callagher (2011) report that in New Zealand, the introduction of an Annual Catch Entitlement (ACE) regime has resulted in an increase in participation by SSF in the inshore fishery. Tolley and Hall-Arber (2015, pp. 401, 402) describe how SSF in New England have fought back against neoliberal fisheries: "Seeking alternatives to the move towards privatisation and corporatisation of the fishing fleet, a new model for organising fishing communities, the Fish Locally Collaborative (FLC) was formed with a goal of…sustaining coastal communities". Tolley and Hall-Arber (2015, p. 402) call the FLC an "alternative economy" to the capitalist model, in that it embraces the notion of the "social economy" and is a not-for-profit social movement organisation which puts moral objectives such as the social justice of reducing material inequality above business objectives. The FLC shows that an alternative to neoliberal policy is possible, but "it is an economy in an increasingly neoliberal environment and thus now requires active political support". As Scholtens et al. (2019, p. 330) point out, the dominant (neo-liberal) discourse needs to be constantly challenged by counter (SSF) discourses: "While dominant discourses typically represent vested interests and may, therefore, be highly resistant to change, alternative- or counter-narratives can be and are developed as well. Such deconstructing, challenging, and reframing of dominant narratives that misrepresent small-scale fishers' interest…requires persistence and strong collaboration with media to affect not only policy but also shape wider public opinion" (see also Witter and Stoll 2017).

7.6 Conclusion

The future of SSF, therefore, lies in our hands: if we want SSF to survive, we must be prepared to support governmental measures, social movements and fishers' associations that help it do so, and "we will make more progress in the long run by appealing to people's hearts than to their wallets" (Douglas McCauley, quoted in Pinkerton 2017, p. 5). We are not recommending that SSF replace industrial fisheries, but only for the inequitable balance that has long existed between them to be redressed so that SSF can continue to exist alongside large-scale fisheries rather than be progressively swallowed up by them. The stakes are high: as an IFCA CFO says, the wider community is at risk of losing its heritage: *'if it collapses…the traditions and tapestry of a little town, coastal village or town will change wholesale'* (KI-106). Another IFCA CFO says *'if these guys aren't making a living and move out of the ports…the ports die as well. So the tourists stop coming because they lose interest and people buy the fishermen's cottages and turn them into holiday homes and then the community starts to shrivel because you've taken away that focal point from the area'* (KI-105). A fisher refers to virtual ghost towns: *'When we lost this industry, we lost lots of other industries as well...So it's really now a really disadvantaged town...this is directly because of the loss of this fishing industry'* (KI-69). Is it too much to ask that steps should be taken to protect the English SSF from extinction in order to safeguard England's coastal heritage? This is an issue that really is too big to ignore.

7.7 Postscript

This 'Big Issue' has assumed even greater urgency with the current threat posed to the English SFF by the Coronavirus, Covid 19. A Welsh inshore shell-fisherman expressed his desperation in an email of 01/04/2020:

> "Three weeks ago all my markets were closed overnight. I had phone calls from merchants, processors saying that they were not buying any shellfish for the foreseeable future as their market were closed. A few days later I had all the restaurants that I supply cancelling orders, as they were closing due to Government advice. I've been informed this week by a shellfish merchant that there is a possibility that the European markets will not be open for six months, possibly a year...This has left me in a dire situation. I'm unable to go out to sea to earn a living as I have nowhere to sell my catch. I have no income coming in, bills to pay, businesses owing me money that have now closed and have no means of paying as they have no cash flow, my most productive months ahead of me with the possibility of no opportunity to fish and sell my catch. Three quarters of my lobster pots are still in storage as there is no market for my shellfish".

References

Abernethy, K., Trebilcock, P., Kebede, B., Allison, E., & Dulvy, N. (2010). Fuelling the decline in UK fishing communities? *ICES Journal of Marine Science, 67*(5), 1076–1085.

Acheson, J. (1988). *The lobster gangs of Maine*. Hanover, NH: University Press of New England.

Acheson, J. (2003). *Capturing the commons: Devising institutions to manage the Maine lobster industry*. Hanover, NH: University Press of New England.

Acott, T., & Urquhart, J. (2014). Sense of -place and socio-cultural values in fishing communities along the English Channel. In J. Urquhart, T. Acott, D. Symes, & M. Zhao (Eds.), *Social issues in sustainable fisheries management* (pp. 257–277). Dordrecht: Springer.

Acott, T., Johnson, D., Stacey, N., & Urquhart. (2018). Reflections of social well-being and the values of small-scale fisheries: Implications for research, policy, and management. In D. Johnson, T. Acott, N. Stacey, & J. Urquhart (Eds.), *Social well-being and the values of small-scale fisheries* (MARE publication series no. 17) (pp. 317–332). Cham: Springer.

Aguilera, S., Cole, J., Finkbeiner, E., et al. (2015). Managing small-scale commercial fisheries for adaptive capacity: Insights from dynamic social-ecological drivers of change in Monterey Bay. *PLoS One, 10*(3), 1–22.

Ahmed, A., & Neelormi, S. (2008). The impacts of climate change on marine socio-ecological systems: The plight of coastal fisher-folk communities in Bangladesh. In *Article presented at conference on coping with global change in marine social ecological systems* (pp. 8–11). Rome: FAO.

AIFCA [Association of Inshore Fisheries and Conservation Authorities]. (2015). *The inshore fisheries and conservation authorities 2011–2015*. Retrieved from www.association-ifca.org.uk

Aizenman, N. (2007, October). Crab processors await decision on guest worker visas. *Washington Post*, p. 13.

Akintola, S., & Fakoya, K. (2017). Small-scale fisheries in the context of traditional post-harvest practice and the quest for food and nutritional security in Nigeria. *Agriculture & Food Security, 6*(1), 34.

Allen, S. (2013, June 18). Recognizing and protecting our small-scale fisheries. *Halifax Media Co-Op*. Retrieved from http://halifax.mediacoop.ca/fr/blog/editor/18297?destination=node%2F18297

Allen, S., & Gough, A. (2006). Monitoring environmental justice impacts: Vietnamese-American longline fishermen adapt to the Hawaii swordfish fishery closure. *Human Organization, 65*(3), 319–328.

Allison, E., & Ellis, F. (2001). The livelihoods approach and management of small-scale fisheries. *Marine Policy, 25*(5), 377–388.

© The Editor(s) (if applicable) and The Author(s), under exclusive license to Springer Nature Switzerland AG 2021

R. Korda et al., *Resilience in the English Small-Scale Fishery*,
https://doi.org/10.1007/978-3-030-54245-0

Allison, E., Horemans, B., & Béné, C. (2006). *Vulnerability reduction and social inclusion: Strategies for reducing poverty among small-scale fisher folk.* Paper presented at the Wetlands, Water and Livelihoods Workshops, Wetland International, St. Lucia, South Africa, 30 January–2 February.

Allison, E., Perry, A., Badjeck, M., Adger, W., Brown, K., Conway, D., & Dulvy, N. (2009). Vulnerability of national economies to the impacts of climate change on fisheries. *Fish and Fisheries, 10*(2), 173–196.

Allison, E., Béné, C., & Andrew, N. (2011). Poverty reduction as a means to enhance resilience in small-scale fisheries. In R. Pomeroy & N. Andrew (Eds.), *Small scale fisheries management: Frameworks and approaches for the developing world* (pp. 216–233). Oxfordshire: CAB International.

Allison, E., Ratner, B., Åsgård, B., Willmann, R., Pomeroy, R., & Kurien, J. (2012). Rights-based fisheries governance: From fishing rights to human rights. *Fish and Fisheries, 13*(1), 14–29.

Amarasinghe, O., & Bavinck, M. (2011). Building resilience: Fisheries cooperatives in southern Sri Lanka. In S. Jentoft & A. Eide (Eds.), *Poverty mosaics: Realities and prospects in small-scale fisheries* (pp. 383–406). Dordrecht: Springer.

Amarasinghe, O., Wanasinghe, W., Jayantha, S., Wishantha Malraj, A., Nishani, D., & Somasiri, A. (2005). Market for fisheries credit: The functioning of the market for fixed capital in fishing communities in Southern Sri Lanka. In *Modernisation and change in marine small-scale fisheries of Southern Sri Lanka* (pp. 181–252). Colombo: Navamaga Printers.

Anbleyth-Evans, J., & Williams, C. (2018). Fishing for justice: England's inshore fisheries social movements and fixed quota allocations. *Human Geography, 11*(1), 28–43.

Anderies, J., Walker, B., & Kinzig, A. (2006). Fifteen weddings and a funeral: Case studies and resilience-based management. *Ecology and Society, 11*(1), 21.

Andrade, H., & Midré, G. (2011). The merits of consensus: Small-scale fisheries as a livelihood buffer in Livingston, Guatemala. In S. Jentoft & A. Eide (Eds.), *Poverty mosaics: Realities and prospects in small-scale fisheries* (pp. 427–448). Dordrecht: Springer.

Andrew, N., & Evans, L. (2011). Approaches and frameworks for management and research in small-scale fisheries. In R. Pomeroy & N. Andrew (Eds.), *Small-scale fisheries management: Frameworks and approaches for the developing world* (pp. 16–34). Oxfordshire: CAB International.

Andrew, N., Béné, C., Hall, S., Allison, E., Heck, S., & Ratner, B. (2007). Diagnosis and management of small-scale fisheries in developing countries. *Fish and Fisheries, 8*(3), 227–240.

Apostle, R., Kasdan, L., & Hanson, A. (1985). Work satisfaction and community attachment among fishermen in Southwest Nova Scotia. *Canadian Journal of Fisheries and Aquatic Sciences, 42*(2), 256–267.

Apostle, R., Barrett, G., Holm, P., Jentoft, S., Mazany, L., McCay, B., & Mikalsen, K. (1998). *Community, state, and market on the North Atlantic rim: Challenges to modernity in the fisheries* (Vol. 4). Toronto: University of Toronto Press.

Appleby, T. (2013). Privatising fishing rights: The way to a fisheries wonderland? *Public Law*, pp. 481–497.

Appleby, T., & Jones, P. (2012). The marine and coastal access act—A hornets' nest? *Marine Policy, 36*(1), 73–77.

Appleby, T., Cardwell, E., Pettipher, J., & Deming, J. (2018). Fishing rights, property rights, human rights: The problem of legal lock-in in UK fisheries. *Elementa: Science of the Anthropocene, 6*(1), 40.

Arias-Schreiber, M., Linke, S., Delaney, A., & Jentoft, S. (2019). Governing the governance: Small-scale fisheries in Europe with focus on the Baltic Sea. In R. Chuenpagdee & S. Jentoft (Eds.), *Transdisciplinarity for small-scale fisheries governance: Analysis and practice* (pp. 357–374). Cham: Springer.

Armitage, D. (2005). Adaptive capacity and community-based natural resource management. *Environmental Management, 35*(6), 703–715.

Armitage, D., Charles, A., & Berkes, F. (2017a). Towards transformative change in the coastal commons. In D. Armitage, A. Charles, & F. Berkes (Eds.), *Governing the coastal commons: Communities, resilience and transformation* (pp. 1–21). London: Routledge.

Armitage, D., Berkes, F., & Charles, A. (2017b). Synthesis: Governing coastal transformations. In D. Armitage, A. Charles, & F. Berkes (Eds.), *Governing the coastal commons: Communities, resilience and transformation* (pp. 252–265). London: Routledge.

Aryeetey, E. (2002). Socio-economic aspects of artisanal marine fisheries management in West Africa. In J. McGlade, P. Cury, K. Koranteng, & N. Hardman-Mountford (Eds.), *The Gulf of Guinea large marine ecosystem* (pp. 323–344). Amsterdam: Elsevier Science.

Badjeck, M., Allison, E., Halls, A., & Dulvy, N. (2010). Impacts of climate variability and change on fishery-based livelihoods. *Marine Policy, 34*(3), 375–383.

Bailey, K. (2018). *Fishing lessons: Artisanal fisheries and the future of our oceans*. London: University of Chicago Press.

Basurto, X., Virdin, J., Smith, H., & Juskus. R. (2017). *Strengthening governance of small-scale fisheries: An initial assessment of theory and practice*. Oak Foundation, Duke University. Retrieved from www.oakfnd.org/environment

Bavinck, M. (2011). Wealth, poverty and immigration: The role of institutions in the fisheries of Tamil Nadu, India. In S. Jentoft & A. Eide (Eds.), *Poverty mosaics: Realities and prospects in small-scale fisheries* (pp. 173–191). Dordrecht: Springer.

Bavinck, M., Berkes, F., Charles, A., Dias, A., Doubleday, N., Nayak, P., & Sowman, M. (2017). The impact of coastal grabbing on community conservation: A global reconnaissance. *Maritime Studies, 16*(8), 1–7.

Bavinck, M., Jentoft, S., & Scholtens, J. (2018). Fisheries as social struggle: A reinvigorated social science research agenda. *Marine Policy, 94*, 46–52.

Beck, U. (1992). *Risk society: Towards a new modernity*. London: Sage.

Beck, T., & Nesmith, C. (2001). Building on poor people's capacities: The case of common property resources in India and West Africa. *World Development, 29*(1), 119–133.

Belhabib, D., Sumaila, U., & Pauly, D. (2015). Feeding the poor: Contribution of West African fisheries to employment and food security. *Ocean and Coastal Management, 111*, 72–81.

Ben Tre Dard. (2009). *Report on performance of the fisheries sector*. Ben Tre Department of Agriculture and Rural Development, Vietnam.

Béné, C. (2003). When fishery rhymes with poverty: A first step beyond the old paradigm on poverty in small-scale fisheries. *World Development, 31*(6), 949–975.

Béné, C. (2006). Small-scale fisheries: Assessing their contribution to rural livelihoods in developing countries. *FAO Fisheries Circular, No*. 1008. Rome: FAO.

Béné, C. (2009a). Are fishers poor or vulnerable? Assessing economic vulnerability in small-scale fishing communities. *Journal of Development Studies, 45*(6), 911–933.

Béné, C. (2009b). Governance and decentralization reforms in small-scale fisheries: An African perspective. In P. Wrammer (Ed.), *Fisheries, sustainability and development* (pp. 253–266). Stockholm: Royal Swedish Academy of Agriculture and Forestry.

Béné, C. (2013). Towards a quantifiable measure of resilience. *IDS Working Papers, 434*, 1–27.

Béné, C., & Friend, R. (2011). Poverty in small-scale fisheries: Old issue, new analysis. *Progress in Development Studies, 11*(2), 119–144.

Béné, C., Bennett, L., & Neiland, A. (2004). Managing small-scale fisheries with reference to poverty. In A. Neiland & C. Béné (Eds.), *Poverty and small-scale fisheries in West Africa* (pp. 83–102). Rome: Kluwer Academic.

Béné, C., Macfadyen, G., & Allison, E. (2007). Increasing the contribution of small-scale fishers to poverty alleviation and food security. *FAO Fisheries Technical Paper, No*. 481. Rome: FAO.

Béné, C., Godfrey-Wood, R., Newsham, A., & Davies, M. (2012). Resilience: New utopia or new tyranny? Reflection about the potentials and limits of the concept of resilience in relation to vulnerability reduction programmes. *IDS Working Paper 405* (p. 61). Brighton: Institute of Development Studies.

Béné, C., Newsham, A., Davies, M., Ulrichs, M., & Godfrey-Wood, R. (2014). Resilience, poverty and development. *Journal of International Development, 26*(5), 598–623.

Béné, C., Devereux, S., & Roelen, K. (2015a). Social protection and sustainable natural resource management: Initial findings and good practices from small-scale fisheries. *FAO Fisheries and Aquaculture Circular No.* 1106. Rome: FAO. 61pp.

Béné, C., Barange, M., Subasinghe, R., Pinstrup-Andersen, P., Merino, G., Hemre, G., & Williams, M. (2015b). Feeding 9 billion by 2050–putting fish back on the menu. *Food Security, 7*(2), 261–274.

Bennett, A. (2017). The influence of neo-liberalization on the success and failure of fishing cooperatives in contemporary small-scale fishing communities: A case study from Yucatán, Mexico. *Marine Policy, 80*, 96–106.

Bennett, N., Govan, H., & Satterfield, T. (2015). Ocean grabbing. *Marine Policy, 57*, 61–68.

Benyon, R. (2013, June 24). Benyon backs small-scale fishermen. *Fishnewseu.com.*

Berkes, F. (2003). Alternatives to conventional management: Lessons from small-scale fisheries. *Environments, 31*(1), 5–20.

Berkes, F. (2007). Community-based conservation in a globalized world. *Proceedings of the National Academy of Sciences, 104*(39), 188–193.

Berkes, F., & Seixas, C. (2005). Building resilience in lagoon social-ecological systems: A local-level perspective. *Ecosystems, 8*(8), 967–974.

Berkes, F., Mahon, R., McConney, P., Pollnac, R., & Pomeroy, R. (2001). *Managing small-scale fisheries: Alternative directions and methods*. Ottowa: International Development Research Centre.

Bhagwati, J. (2004). Anti-globalization: Why? *Journal of Policy Modelling, 26*, 439–463.

Bleiker, R. (2008). Politics after Seattle: Dilemmas of the anti-globalization movement. *Cultures & Conflits*, 1–19. Retrieved from http://conflits.revues.org/1057.

Blount, B. (2007). Life on the water: A historical cultural model of African American fishermen on the Georgia coast (USA). *NAPA Bulletin, 28*(1), 109–122.

Bodwitch, H. (2017). Challenges for New Zealand's individual transferable quota system: Processor consolidation, fisher exclusion, and Māori quota rights. *Marine Policy, 80*, 88–95.

Bown, N., Gray, T., & Stead, S. (2013). *Contested forms of governance in marine protected areas: A study of co-management and adaptive co-management*. London: Routledge.

Boyle, D. (2009). *Localism: Unravelling the supplicant state*. London: New Economics Foundation. 24pp.

Brand, F., & Jax, K. (2007). Focusing on the meaning(s) of resilience: Resilience as a descriptive concept and a boundary object. *Ecology and Society, 12*(1), 23–38.

Brattland, C. (2014). A cybernetic future for small-scale fisheries. *Maritime Studies, 13*(18), 1–21.

Britton, E. (2012a, June 25). Giving small-scale fisheries a place: The knowledge and governance challenges. *MARE Policy Day*. Amsterdam. 18pp.

Britton, E. (2012b). Women as agents of wellbeing in Northern Ireland's fishing households. *Maritime Studies, 11*(16), 1–22.

Brookfield, K., Gray, T., & Hatchard, J. (2005). The concept of fisheries-dependent communities: A comparative analysis of four UK case studies: Shetland, Peterhead, North Shields and Lowestoft. *Fisheries Research, 72*(1), 55–69.

Brundtland Report. (1987). *Our common future*. New York: World Commission on Environment and Development.

Campling, L., & Havice, E. (2014). The problem of property in industrial fisheries. *Journal of Peasant Studies, 41*(5), 707–727.

Campling, L., Havice, E., & Howard, P. (2012). The political economy and ecology of capture fisheries: Market dynamics, resource access, and relations of exploitation and resistance. *Journal of Agrarian Change, 12*(2–3), 177–203.

Cappell, R. (n.d.). *Project inshore: Stage 3—Strategic sustainability review: Isles of Scilly IFCA*. Edinburgh: Accoura.

Carbonetti, B., Pomeroy, R., & Richards, D. (2014). Overcoming the lack of political will in small-scale fisheries. *Marine Policy, 44*, 295–301.

Cardwell, E. (2012). Invisible fishermen: The rise and fall of the British small boat fleet. In P. Højrup & K. Schriewer (Eds.), *European fisheries at a tipping point* (pp. 367–412). Murcia: Universidad de Murcia.

Carpenter, G. (2017). *Not in the same boat: The economic impact of Brexit across UK fishing fleets*. London: New Economics Foundation. 115pp.

Carvalho, N., Edwards-Jones, G., & Isidro, E. (2011). Defining scale in fisheries: Small versus large-scale fishing operations in the Azores. *Fisheries Research, 109*(2–3), 360–369.

Catchpole, T., & Revill, A. (2008). Gear technology in *Nephrops* trawl fisheries. *Reviews in Fish Biology and Fisheries, 17*(4), 893–938.

Catchpole, T., Ribiero-Santos, A., Mangi, S., & Hedley, C. (2017). The challenges of the landing obligation in EU fisheries. *Marine Policy, 62*, 78–86.

Cernea, M., & Schmidt-Soltau, K. (2006). Poverty risks and national parks: Policy issues in conservation and resettlement. *World Development, 34*(10), 1808–1830.

Chambers, R. (1989). Editorial introduction: Vulnerability, coping and policy. *IDS Bulletin, 20*(2), 1–7.

Chambers, C., & Carothers, C. (2017). Thirty years after privatization: A survey of Icelandic small-boat fishermen. *Marine Policy, 80*, 69–80.

Chandler, D. (2014). *Resilience: The governance of complexity*. London: Routledge.

Chandler, D., & Reid, J. (2016). *The neo-liberal subject: Resilience, adaptation and vulnerability*. London: Rowman & Littlefield.

Charles, A. (2011a). Good practices for governance of small-scale fisheries. In R. Chuenpagdee (Ed.), *World small-scale fisheries: Contemporary visions* (pp. 285–298). Delft: Eburon.

Charles, A. (2011b). Human rights and fishery rights in small-scale fisheries management. In R. Pomeroy & N. Andrew (Eds.), *Small scale fisheries management: Frameworks and approaches for the developing world* (pp. 59–74). Oxfordshire: CAB International.

Charles, A. (2013). Governance of tenure in small-scale fisheries: Key considerations. *Land Tenure Journal, 1*(13), 9–37.

Charmaz, K. (2006). *Constructing grounded theory: A practical guide through qualitative analysis*. London: Sage.

Cheung, W., Lam, V., Sarmiento, J., Kearney, K., Watson, R., Zeller, D., & Pauly, D. (2010). Large-scale redistribution of maximum fisheries catch potential in the global ocean under climate change. *Global Change Biology, 16*(1), 24–35.

Chuenpagdee, R. (2011a). Preface. In R. Chuenpagdee (Ed.), *World small-scale fisheries: Contemporary visions* (pp. 9–14). Delft: Eburon.

Chuenpagdee, R. (2011b). Too big to ignore: Global research network for the future of small-scale fisheries. In R. Chuenpagdee (Ed.), *World small-scale fisheries: Contemporary visions* (pp. 383–394). Delft: Eburon.

Chuenpagdee, R. (2019). Too big to ignore—A transdisciplinary journey. In R. Chuenpagdee & S. Jentoft (Eds.), *Transdisciplinarity for small-scale fisheries governance: Analysis and practice* (pp. 15–31). Cham: Springer.

Chuenpagdee, R., & Jentoft, S. (2011). Situating poverty: A chain analysis of small-scale fisheries. In S. Jentoft & A. Eide (Eds.), *Poverty mosaics: Realities and prospects in small-scale fisheries* (pp. 27–42). Dordrecht: Springer.

Chuenpagdee, R., & Jentoft, S. (2015). Exploring challenges in small-scale fisheries governance. In S. Jentoft & R. Chuenpagdee (Eds.), *Interactive governance for small-scale fisheries* (pp. 3–16). Cham: Springer.

Chuenpagdee, R., & Juntarashote, K. (2011). Learning from the experts: Attaining sufficiency in small-scale fishing communities in Thailand. In S. Jentoft & A. Eide (Eds.), *Poverty mosaics: Realities and prospects in small-scale fisheries* (pp. 309–331). Dordecht: Springer.

Chuenpagdee, R., Degnbol, P., Bavinck, M., Jentoft, S., Johnson, D., Pullin, R., & Williams, S. (2005). Challenges and concerns in capture fisheries and aquaculture. In *Centre for maritime*

research. Fish for life: Interactive governance for fisheries (no. 3) (pp. 25–40). Amsterdam: Amsterdam University Press.
Chuenpagdee, R., Liguori, L., Palomares, M., & Pauly, D. (2006). Bottom-up, global estimates of small-scale marine fisheries catches. *Fisheries Centre Research Report, 14*(8), 110.
Cinner, J., Daw, T., & McClanahan, T. (2009). Socioeconomic factors that affect artisanal fishers' readiness to exit a declining fishery. *Conservation Biology, 23*(1), 124–130.
Clark, C. (2006). *The worldwide crisis in fisheries: Economic models and human behaviour*. Cambridge: Cambridge University Press.
Clay, P., & Olson, J. (2008). Defining "fishing communities": Vulnerability and the Magnuson-Stevens fishery conservation and management act. *Human Ecology Review, 15*(2), 143–160.
Coaffee, J. (2005). New localism and the management of regeneration. *International Journal of Public Sector Management, 18*(2), 108–113.
Coaffee, J., & Johnston, L. (2005). The management of local government modernization: Area decentralization and pragmatic localism. *International Journal of Public Sector Management, 18*(2), 164–177.
Cohen, P., Allison, E., Andrew, N., et al. (2019). Securing a just space for small-scale fisheries in the blue economy. *Frontiers in Marine Science, 6*, 171.
Cole, C. (1996). Can we resolve uncertainty in marine fisheries management? In J. Lemon (Ed.), *Scientific uncertainty and environmental problem solving* (pp. 233–263). Oxford: Blackwell Science.
Coulthard, S. (2012). Can we be both resilient and well, and what choices do people have? Incorporating agency into the resilience debate from a fisheries perspective. *Ecology and Society, 17*(1), 12.
Coulthard, S., & Britton, E. (2015). Waving or drowning: An exploration of adaptive strategies amongst fishing households and implications for well-being outcomes. *Sociologia Ruralis, 55*(3), 275–290.
Courtney, C., Jhaveri, N., Pomeroy, R., & Brooks, S. (2016). *Marine tenure and small-scale fisheries: Learning from the Philippines experience*. Washington, DC: USAID Tenure and Global Climate Change Program.
CRFM [Caribbean Regional Fisheries Mechanism]. (2018, May 23). Caribbean fisheries ministers endorse protocol for securing small-scale fisheries. *Press Release*. Belize City.
Crilly, R., & Esteban, A. (2013). Small versus large-scale, multi-fleet fisheries: The case for economic, social and environmental access criteria in European fisheries. *Marine Policy, 37*, 20–27.
Crosoer, D., Van Sittert, L., & Ponte, S. (2006). The integration of South African fisheries into the global economy: Past, present and future. *Marine Policy, 30*(1), 18–29.
Curtis, F. (2003). Eco-localism and sustainability. *Ecological Economics, 22*(9), 83–102.
D'Armengol, L., Castillo, M., Ruiz-Mallen, I., & Corbera, E. (2018). A systematic review of co-managed small-scale fisheries: Social diversity and adaptive management improve outcomes. *Global Environmental Change, 52*, 212–225.
Daboub, A., & Calton, J. (2002, April 11–12). World trade and anti-globalization: Ethical implications. In: *Seventh Annual Conference: Free trade and the western hemisphere: The challenges and the future*. Texas: Texas A&M International University.
DAFF [Department of Agriculture, Forestry and Fisheries]. (2012). Policy for the small-scale fisheries sector in South Africa. *Government Gazette*, 564(35455). Pretoria: Republic of South Africa. 58pp.
Davies, P., Williams, C., Carpenter, G., & Stewart, B. (2018). Does size matter? Assessing the use of vessel length to manage fisheries in England. *Marine Policy, 97*, 202–210.
Davis, R. (2015). 'All in': Snow crab, capitalization and the future of small-scale fisheries in Newfoundland. *Marine Policy, 61*, 323–330.
Davis, A., & Ruddle, K. (2012). Massaging the misery: Recent approaches to fisheries governance and the betrayal of small-scale fisheries. *Human Organization, 71*(3), 244–254.

Davis, A., & Wagner, J. (2006). A right to fish for a living? The case for coastal fishing people's determination of access and participation. *Ocean and Coastal Management, 49*(7–8), 476–497.

Davoudi, S. (2012). Resilience: A bridging concept or a dead end? *Planning Theory and Practice, 13*(2), 299–333.

DCLG [Department for Communities and Local Government]. (2011). *A plain English guide to the Localism Act*. London: Author.

De Mattos, S., & Wojciechowski, M. (2019). Existing institutional and legal framework for small-scale fisheries development in Brazil. In S. Salas, M. Barragán-Paladines, & R. Chuenpagdee (Eds.), *Viability and sustainability of small-scale fisheries in Latin America and the Caribbean* (pp. 495–511). Cham: Springer.

De Schutter, O. (2012, November 27). How development organisations can tackle the fisheries challenge. *The Guardian*.

de Vos, B., & Kraan, M. (2015). To define or not to define: Implications for the governability of small-scale coastal fisheries in the Netherlands. In S. Jentoft & R. Chuenpagdee (Eds.), *Interactive governance for small-scale fisheries. Global reflections* (pp. 629–647). Dordrecht: Springer.

Defeo, O., Castrajón, M., Peréz-Castañeda, R., Castilla, J., Gutiérrez, N., Esssington, T., & Folke, C. (2016). Co-management in Latin American small-scale shellfisheries: Assessment for long-term case studies. *Fish and Fisheries, 17*, 176–192.

Defra. (2008). *The English inshore fleet—looking to the future consultation*. Retrieved from https://www.legislation.gov.uk/ukia/2008/266/pdfs/ukia_20080266_en.pdf

Defra. (2009). *Marine and coastal access bill: Policy paper*. Retrieved from http://www.defra.gov.uk/environment/marine/documents/legislation/mab-policy.pdf

Defra. (2012). *Report of the habitats and wild birds directives implementation review*. London: Crown copyright. Retrieved from https://assets.publishing.service.gov.uk/government/uploads/system/uploads/attachment_data/file/69513/pb13724-habitats-review-report.pdf

Defra. (2014). *Triennial review of the Marine Management Organization*. London.

Defra. (2015). *Inshore Fisheries and Conservation Authorities: Conduct and operation 2010–2014*. London.

Defra. (2018). *Sustainable fisheries for future generations. Cm 9660*. London.

Degnbol, P., Gislason, H., Hanna, S., Jentoft, S., Raakjaer Nielsen, J., Sverdrup-Jensen, S., & Wilson, D. (2006). Painting the floor with a hammer: Technical fixes in fisheries management. *Marine Policy, 30*(5), 534–543.

Denscombe, M. (2003). *The good research guide for small-scale social research projects*. Buckingham: McGraw-Hill Education.

Denzin, N., & Lincoln, Y. (2008). *Collecting and interpreting qualitative materials* (Vol. 3). London: Sage.

Deswandi, R., Glaser, M., Fersen, S., Hornidge, A., & Antweiler, C. (2012). What makes a social system resilient? Two fishing communities in Indonesia. In A. Hornidge & C. Antweiler (Eds.), *Environmental uncertainty and local knowledge: Southeast Asia as a laboratory of global ecological change*. Bonn: University of Bonn.

Doumbouya, A., Camara, O., Mamie, J., Intchama, J., Jarra, A., Ceesay, S., & Belhabib, D. (2017). Assessing the effectiveness of monitoring control and surveillance of illegal fishing: The case of West Africa. *Frontiers in Marine Science, 4*, 50.

Dudley, N. (2008). *Guidelines for applying protected area management categories*. Gland: IUCN.

Duggan, D., & Kochen, M. (2016). Small in scale but big in potential: Opportunities and challenges for fisheries certification of Indonesian small-scale tuna fisheries. *Marine Policy, 67*, 30–39.

Dulvy, N., Rogers, S., Jennings, S., Stelzenmüller, V., Dye, S., & Skjoldal, H. (2008). Climate change and deepening of the North Sea fish assemblage: A biotic indicator of warming seas. *Journal of Applied Ecology, 45*(4), 1029–1039.

Edwards, D., & Edwards, D. (2017). Licence banks as a tool to mitigate corporate control of fisheries: A British Columbia groundfish example. *Marine Policy, 80*, 141–146.

Edwards, D., & Pinkerton, E. (2020). Priced out of ownership: Quota leasing impacts on the financial performance of owner-operators. *Marine Policy, 111*, 103718.

Eide, A., Bavinck, M., & Raakjær, J. (2011). Avoiding poverty: Distributing wealth in fisheries. In S. Jentoft & A. Eide (Eds.), *Poverty mosaics: Realities and prospects in small-scale fisheries* (pp. 13–25). Dordrecht: Springer.

EU Commission. (2009). *Green paper: Reform of the common fisheries policy, CON (2009) 163 final*. Brussels: European Commission.

EU Commission. (2018). *Ministerial Declaration on increased opportunities in the Mediterranean and Black Sea*. Valletta: EU Commission.

EU Parliament. (2016, June 27). *Local and regional leaders call for special provisions to support small-scale coastal fisheries*. Meeting of the EU Parliament's Committee of the Regions.

Evans, L., & Andrew, N. (2011). Diagnosis and the management constituency of small-scale fisheries. In R. Pomeroy & N. Andrew (Eds.), *Small-scale fisheries management: Frameworks and approaches for the developing world* (pp. 35–58). Oxfordshire: CAB International.

FAO. (1995). *Code of conduct for responsible fisheries*. Rome.

FAO. (1999). *Number of fishers 1970–1996*. Fishery Information Data and Statistics Unit. Rome: FAO Fisheries.

FAO. (2000). *Coastal fishing communities in Thailand*. Rome: FAO Fisheries.

FAO. (2003). *Report of the second session of the working party on small-scale fisheries*. Report No. 735. Advisory Committee on Fisheries Research. Bangkok: FAO Fisheries.

FAO. (2005). *Increasing the contribution of small-scale fisheries to poverty alleviation and food security*. FAO Technical Guidelines for Responsible Fisheries No. 10. Rome.

FAO. (2007). *Increasing the contribution of small-scale fisheries to poverty alleviation and food security*. FAO Fisheries Technical Paper 481. Rome.

FAO. (2008). *Global conference on small-scale fisheries: Securing sustainable small-scale fisheries: Bringing together responsible fisheries and social development*. Rome: FAO.

FAO. (2015). *Voluntary guidelines for securing sustainable small-scale fisheries in the context of food security and poverty eradication*. Rome.

Faraco, L., Andriguetto-Filho, J., Daw, T., Lana, P., & Teixeira, C. (2016). Vulnerability among fishers in southern Brazil and its relation to marine protected areas in a scenario of declining fisheries. *Desenvolvimento e Meio Ambiente, 38*, 51–76.

Ferdman, A. (2017). Why the intrinsic value of public goods matters. *Critical Review of International Social and Political Philosophy, 21*(5), 661–676.

Finkbeiner, E., & Basurto, X. (2015). Re-defining co-management to facilitate small-scale fisheries reform: An illustration from Northwest Mexico. *Marine Policy, 51*, 433–441.

Finkbeiner, E., Ayers, A., Kittinger, J., & Crowder, L. (2015). A comparison of small-scale fisheries governability: Baja California Sur, Mexico and the Hawaiian Islands. In S. Jentoft & R. Chuenpagdee (Eds.), *Interactive governance for small-scale fisheries* (pp. 199–221). Cham: Springer.

Fish Site. (2013, June 11). Sustainability map of England's 450+ inshore fisheries completed. *The Fish Site*.

Fishnews. (2015, March 30). Seafish and MSC unveil first of their 'roadmaps' for sustainable fisheries throughout England. *Fishnews.com*.

Fleet Register. (2016). *Advanced search: Fleet register*. Retrieved from http://ec.europa.eu/fisheries/fleet/index.cfm

Foley, P., Mather, C., & Neiss, B. (2015). Governing enclosure for coastal communities: Social embeddedness in a Canadian shrimp fishery. *Marine Policy, 61*, 390–400.

Folke, C., Carpenter, S., Elmqvist, T., Gunderson, L., Holling, C., & Walker, B. (2002). Resilience and sustainable development: Building adaptive capacity in a world of transformations. *AMBIO, 31*(5), 437–440.

Folke, C., Colding, J., & Berkes, F. (2003). Synthesis: Building resilience and adaptive capacity in social-ecological systems. In F. Berkes, J. Colding, & C. Folke (Eds.), *Navigating social-ecological systems: Building resilience for complexity and change* (pp. 352–387). Cambridge: Cambridge University Press.

Folke, C., Carpenter, S., Walker, B., Scheffer, M., Chapin, T., & Rockström, J. (2010). Resilience thinking: Integrating resilience, adaptability and transformability. *Ecology and Society, 15*(4), 20.

Foundational Economy. (2018). *Introducing the foundational economy*. Retrieved from https://foundationaleconomy.com/introduction

Francis, J., Johnston, M., Robertson, C., Glidewell, L., Entwistle, V., Eccles, M., & Grimshaw, J. (2010). What is an adequate sample size? Operationalising data saturation for theory-based interview studies. *Psychology and Health, 25*(10), 1229–1245.

Franz, N., & Barragán-Paladines, M. (2017). The small-scale fisheries guidelines: A first account of developments since their endorsement in 2014. In S. Jentoft, R. Chuenpagdee, M. Barragán-Paladines, & N. Franz (Eds.), *The small-scale fisheries guidelines: Global implementation* (pp. 35–54). Dordrecht: Springer.

Fröcklin, S., de la Torre-Castro, M., Lindström, L., & Jiddawi, N. (2013). Fish traders as key actors in fisheries: Gender and adaptive management. *AMBIO, 42*(8), 951–962.

García-Flórez, L., Morales, J., Gaspar, M., Castilla, D., Mugerza, E., Berthou, P., & Arregi, L. (2014). A novel and simple approach to define artisanal fisheries in Europe. *Marine Policy, 44*, 152–159.

Gasalla, M. (2011). Do all answers lie within (the community)? Fishing rights and marine conservation. In R. Chuenpagdee (Ed.), *World small-scale fisheries: Contemporary visions* (pp. 185–203). Delft: Eburon.

Gatewood, J., & McCay, B. (1988). Job satisfaction and the culture of fishing: A comparison of six New Jersey Fisheries. *Maritime Anthropological Studies, 1*(2), 103–128.

Gelcich, S., Edward-Jones, G., Kaiser, M., & Castilla, J. (2006). Co-management policy can reduce resilience in traditionally managed marine ecosystems. *Ecosystems, 9*, 951–966.

Glantz, M., & Thompson, J. (1981). *Resource management and environmental uncertainty; lessons from coastal upwelling fisheries*. Chichester: Wiley.

Glaser, B., & Strauss, A. (1967). *The discovery of grounded theory: Strategies for qualitative research*. New York: Academic Press.

Gómez, S., Lloret, J., Demestre, M., & Riera, V. (2006). The decline of the artisanal fisheries in Mediterranean coastal areas: The case of Cap de Creus (Cape Creus). *Coastal Management, 34*(2), 217–232.

González, M. (2011). To make a fishing life: Community empowerment in small-scale fisheries in the Pearl Lagoon, Nicaragua. In S. Jentoft & A. Eide (Eds.), *Poverty mosaics: Realities and prospects in small-scale fisheries* (pp. 275–308). Dordrecht: Springer.

Gordon, S. (1954). Theory of a common-property resource: The fishery. *Journal of Political Economy, 62*(2), 124–142.

Gordon, D., & Munro, G. (1996). *Fisheries and uncertainty: A precautionary approach to resource management*. Calgary: University of Calgary Press.

Gray, T., Korda, R., Stead, S., & Jones, E. (2011). Quota discarding and distributive justice: The case of the under-10 m fishing fleet in Sussex, England. *Marine Policy, 35*(2), 122–129.

Greenpeace. (2013, January). A wolf in shrimp's clothing. Exposing the national federation of fishermen's organisation. *Ocean Inquirer*, Issue 05.

Gubrium, J., & Holstein, J. (2002). *Handbook of interview research: Context and method*. Thousand Oaks, CA: Sage Publications.

Gudeman, S. (2001). *The anthropology of economy*. Oxford: Blackwell.

Guillen, J., Holmes, S., Carvalho, N., Casey, J., Dorner, H., Gibin, M., Mannini, A., Vasilakopoulos, P., & Zanzi, A. (2018). A review of the European Union landing obligation focusing on its implications for fisheries and the environment. *Sustainability, 10*(4), 900.

Gunderson, L., & Holling, C. (Eds.). (2002). *Panarchy: Understanding transformations in human and natural systems*. Washington: Island Press.

Gustavsson, M., & Riley, M. (2018). Women, capitals and fishing lives: Exploring gendered dynamics in the Llŷn Peninsula small-scale fishery (Wales, UK). *Maritime Studies, 17*, 223–231.

Haakonsen, J. (1992). Industrial vs. artisanal fisheries in West Africa: The lessons to be learnt. In I. Tvedten & B. Hersoug (Eds.), *Fishing for development: Small-scale fisheries in Africa* (pp. 33–53). Uppsala: Scandinavian Institute of African Studies.

Habib, A. (2001). *Delipara: An obscure fishing village of Bangladesh*. Raipur: Community Development Centre.

Hall, S. (2011). Climate change and other external drivers in small-scale fisheries: Practical steps for responding. In R. Pomeroy & N. Andrew (Eds.), *Small scale fisheries management: Frameworks and approaches for the developing world* (pp. 132–159). Cambridge: CAB International.

Haque, C., Idrobo, C., Berkes, F., & Giesbrecht, D. (2015). Small-scale fishers' adaptations to change: The role of formal and informal credit in Paraty, Brazil. *Marine Policy, 51*, 401–407.

Hardin, G. (1968). The tragedy of the commons. *Science, 162*(3859), 1243–1248.

Hatcher, A. (1997). Producers' organisations and devolved fisheries management in the United Kingdom: Collective and individual quota systems. *Marine Policy, 21*(6), 519–533.

Hatcher, A., & Gordon, D. (2005). Further investigations into the factors affecting compliance with UK fishing quotas. *Land Economics, 81*(1), 71–86.

Hatcher, A., & Read, A. (2001). The allocation of fishing rights in UK fisheries. In *Case studies on the allocation of transferable quota rights in fisheries*. Rome: FAO.

Hatcher, A., Pascoe, S., Banks, R., & Arnason, R. (2002). *Future options for UK fish quota management*. Portsmouth: University of Portsmouth.

Hauck, M. (2008). Rethinking small-scale fisheries compliance. *Marine Policy, 32*(4), 635–642.

Held, D., & McGrew, A. (2003). The great globalization debate: An introduction. In D. Held & A. McGrew (Eds.), *The global transformations reader: An introduction to the globalization debate* (pp. 1–50). Cambridge: Polity Press.

Hennink, M., Hutter, I., & Bailey, A. (2011). *Qualitative research methods*. London: Sage.

Henry, A., & Johnson, T. (2015). Understanding social resilience in the Maine lobster industry. *Marine and Coastal Fisheries, 7*(1), 33–43.

Hines, C. (2000). *Localization: A global manifesto*. London: Earthscan.

Hinrichs, C. (2003). The practice and politics of food system localization. *Journal of Rural Studies, 19*, 33–45.

Hjul, J. (2015, March 30). Seafish launches inshore fishing road map. *FishUpdate.com*.

Hoffmann, S. (2003). Clash of globalizations. In D. Held & A. McGrew (Eds.), *The global transformations reader: An introduction to the globalization debate* (pp. 106–111). Cambridge: Polity Press.

Holcombe, R. (2000). Public goods theory and public policy. *Journal of Value Inquiry, 34*, 273–286.

Holling, C., & Gunderson, L. (2002). Resilience and adaptive cycles. In L. Gunderson & C. Holling (Eds.), *Panarchy: Understanding transformations in human and natural systems* (pp. 25–62). Washington, DC: Island Press.

Horemans, B. (1998). *The state of artisanal fisheries in West Africa in 1997*. IDAF/DIPA Technical Report No. 84.

Hosch, G. (2002, March). Don't jump the fence. Fisheries development and management must choose an integrated approach. *Samudra*.

Høst, J. (2015). Governing through markets: Societal objectives, private property rights and small-scale fisheries in Denmark. In S. Jentoft & R. Chuenpagdee (Eds.), *Interactive governance for small-scale fisheries. Global reflections* (pp. 319–336). Dordrecht: Springer.

Huggins, C. (2018). *Government's fisheries white paper: What it does and doesn't say*. Brexit Research: The UK in a changing Europe. Retrieved from http://ukandeu.ac.uk/analysis/

Ifremer. (2007). *Small-scale coastal fisheries in Europe*, Final report of the contract no. Fish/2005/10. 447pp.

Intchama, J., Belhabib, D., Jumpe, T., & Joaquim, R. (2018). Assessing Guinea Bissau's legal and illegal unreported and unregulated fisheries and the surveillance efforts to tackle them. *Frontiers in Marine Science, 5*, 79.

Isaacs, M. (2006). Small-scale fisheries reform: Expectations, hopes and dreams of 'a better life for all'. *Marine Policy, 30*, 51–59.

Isaacs, M. (2015). The governability of small scale fisheries food system in South Africa—The case of snoek and west coast rock lobster. In S. Jentoft & R. Chuenpagdee (Eds.), *Interactive governance for small-scale fisheries—Global reflections* (pp. 245–262). Dordrecht: Springer.

Islam, M. (2011). Living on the margin: The poverty-vulnerability nexus in the small-scale fisheries of Bangladesh. In S. Jentoft & A. Eide (Eds.), *Poverty mosaics: Realities and prospects in small-scale fisheries* (pp. 71–95). Dordrecht: Springer.

Islam, D., & Berkes, F. (2016). Can small-scale commercial and subsistence fisheries co-exist? Lessons from an indigenous community in northern Manitoba, Canada. *Maritime Studies, 15*(1), 1–16.

Jacinto, E., & Pomeroy, R. (2011). Developing markets for small-scale fisheries: Utilizing the value chain approach. In R. Pomeroy & N. Andrew (Eds.), *Small-scale fish management framework approaches in the developing world* (pp. 160–177). Wallingford: CAB International.

Jacob, S., Farmer, F., Jepson, M., & Adams, C. (2001). Landing a definition of fishing dependent communities: Potential social science contributions to meeting National Standard 8. *Fisheries, 26*(10), 16–22.

Jacobsen, R. (2013). Small-scale fisheries in Greenlandic planning—The becoming of a governance problem. *Maritime Studies, 12*(2), 1–19.

Jacquet, J., & Pauly, D. (2008). Funding priorities: Big barriers to small-scale fisheries. *Conservation Biology, 22*(4), 832–835.

Jameson, H. (2013). What is localism? *Municipal year book 2013*.

Jentoft, S. (2000). Legitimacy and disappointment in fisheries management. *Marine Policy, 24*(2), 141–148.

Jentoft, S. (2011). Fisheries development as freedom. In R. Chuenpagdee (Ed.), *World small-scale fisheries: Contemporary visions* (pp. 367–379). Delft: Eburon.

Jentoft, S. (2019). Governing change in small-scale fisheries: Theories and assumptions. In R. Chuenpagdee & S. Jentoft (Eds.), *Transdisciplinarity for small-scale fisheries governance: Analysis and practice* (pp. 305–320). Cham: Springer.

Jentoft, S., & Chuenpagdee, R. (2015a). Assessing governability of small-scale fisheries. In S. Jentoft & R. Chuenpagdee (Eds.), *Interactive governance for small-scale fisheries: Global reflections* (pp. 17–35). Dordrecht: Springer.

Jentoft, S., & Chuenpagdee, R. (2015b). Enhancing the governability of small-scale fisheries through interactive governance. In S. Jentoft & R. Chuenpagdee (Eds.), *Interactive governance for small-scale fisheries: Global reflections* (pp. 727–747). Dordrecht: Springer.

Jentoft, S., & Chuenpagdee, R. (2017). From rhetoric to reality: Implementing the voluntary guidelines for securing sustainable small-scale fisheries. In S. Jentoft, R. Chuenpagdee, M. Barragan-Paladines, & N. Franz (Eds.), *The small-scale fisheries guidelines: Global implementation* (pp. 789–812). Dordrecht: Springer.

Jentoft, S., & Chuenpagdee, R. (2018). From poverty to well-being in small-scale fisheries: The governability challenge. In D. Johnson, T. Acott, N. Stacey, & J. Urquhart (Eds.), *Social well-being and the values of small-scale fisheries* (MARE publication series no. 17) (pp. 293–315). Cham: Springer.

Jentoft, S., & Eide, A. (2011). Setting the stage. In S. Jentoft & A. Eide (Eds.), *Poverty mosaics: Realities and prospects in small-scale fisheries* (pp. 1–10). Dordrecht: Springer.

Jentoft, S., & Johnsen, J. (2015). The dynamics of small-scale fisheries in Norway: From adaptamentality to governability. In S. Jentoft & R. Chuenpagdee (Eds.), *Interactive governance for small-scale fisheries. Global reflections* (pp. 705–723). Dordrecht: Springer.

Jentoft, S., & Midré, G. (2011). The meaning of poverty: Conceptual issues in small-scale fisheries research. In S. Jentoft & A. Eide (Eds.), *Poverty mosaics: Realities and prospects in small-scale fisheries* (pp. 43–68). Dordrecht: Springer.

Jentoft, S., Onyango, P., & Islam, M. (2010). Freedom and poverty in the fishery commons. *International Journal of the Commons, 4*(1), 345–366.

Jentoft, S., Eide, A., Bavinck, M., Chuenpagadee, R., & Raakjær, J. (2011). A better future: Prospects for small-scale fishing people. In S. Jentoft & A. Eide (Eds.), *Poverty mosaics: Realities and prospects in small-scale fisheries* (pp. 451–469). Dordrecht: Springer.

Jentoft, S., Chuenpagdee, R., Barragan-Paladines, M., & Franz, N. (Eds.). (2017a). *The small-scale fisheries guidelines: Global implementation*. Dordrecht: Springer.

Jentoft, S., Chuenpagdee, R., Franz, N., & Barragan-Paladines, M. (2017b). Implementing the voluntary guidelines for securing small-scale fisheries. In S. Jentoft, R. Chuenpagdee, M. Barragan-Paladines, & N. Franz (Eds.), *The small-scale fisheries guidelines: Global implementation* (pp. 3–13). Dordrecht: Springer.

Johannes, R. (1982). Traditional conservation methods and protected marine areas in Oceania. *AMBIO, 11*(5), 258–261.

Jóhannesson, G., Skaptadóttir, U., & Benediktsson, K. (2003). Coping with social capital? The cultural economy of tourism in the north. *Sociologia Ruralis, 43*(1), 3–16.

Johnson, D. (2006). Category, narrative and value in the governance of small-scale fisheries. *Marine Policy, 30*, 747–756.

Johnson, J., & Orbach, M. (1990). Migratory fishermen: A case study in inter-jurisdictional natural resource management. *Ocean and Shoreline Management, 13*(3–4), 231–252.

Johnson, D., & Pálsson, S. (2015). Governability and its discontents in the fishery of Lake Winnipeg since the late 1960s: The view from Gimli. In S. Jentoft & R. Chuenpagdee (Eds.), *Interactive governance for small-scale fisheries. Global reflections* (pp. 281–298). Dordrecht: Springer.

Johnson, T., Henry, A., & Thompson, C. (2014). Qualitative indicators of social resilience in small-scale fishing communities: An emphasis on perceptions and practice. *Human Ecology Review, 20*(2), 97–115.

Jones, P. (2008). Fishing industry and related perspectives on the issues raised by no-take marine protected area proposals. *Marine Policy, 32*(4), 749–758.

Jones, E., Gray, T., & Umponstira, C. (2010). Small-scale fishing: Perceptions and threats to conserving a livelihood in the province of Phang-Nga, Thailand. *Environment Asia, 3*(1), 1–7.

Jul-Larsen, E. (1992). Attitudes towards modernization in African small-scale fisheries. In I. Tvedten & B. Hersoug (Eds.), *Fishing for development: Small-scale fisheries in Africa* (pp. 70–90). Uppsala: Scandinavian Institute of African Studies.

Kaiser, M. (2019, October 17). Beginning of a process. *Fishing News*, p. 2.

Kallhoff, A. (2014). Why societies need public goods. *Critical Review of International Social and Political Philosophy, 17*(6), 635–651.

Kaplan, I. (1988). Women who go to sea: Working in the commercial fishing industry. *Journal of Contemporary Ethnography, 16*(4), 491–514.

Katz, B., & Nowak, J. (2018). *The new localism: How cities can thrive in the age of populism*. Washington, DC: Brookings Institution Press.

Kelleher, C. (2005). *Discards in the world's marine fisheries: An update*. FAO Fisheries Technical Paper 470. Rome: FAO.

Khan, A. (1998). *Poverty in China in the period of globalization: New evidence on trend and pattern*. Issues in Development Discussion Paper 22. Geneva: International Labor Organization.

Kitner, K. (2006). Beeliners, pinkies, and kitties: Mobility and marginalization in the South Atlantic snapper grouper fishery. *Human Organization, 65*(3), 294–306.

Kleiber, D., Harris, L., & Vincent, A. (2014). Gender and small-scale fisheries: A case for counting women and beyond. *Fish and Fisheries, 16*, 547–562.

Knott, C., & Neiss, B. (2017). Privatization, financialization and ocean grabbing in New Brunswick herring fisheries and salmon aquaculture. *Marine Policy, 80*, 10–18.

Knutson, P. (2017). Escaping the corporate net: Pragmatics of small-boat direct marketing in the U.S. salmon fishing industry of the North-Eastern Pacific. *Marine Policy, 80*, 123–129.

Kochalski, A. (2017). Sustainability and conflict in small-scale fisheries (Doctoral dissertation, University of Liverpool).

Kohls, R., & Uhl, J. (2001). *Marketing of agricultural products (no. 7)*. London: Prentice Hall.

Kolding, J., & van Zwieten, P. (2011). The tragedy of our legacy: How do global management discourses affect small-scale fisheries in the South? *Forum for Development Studies, 38*(3), 267–297.

Kolding, J., Béné, C., & Bavinck, M. (2014). Small-scale fisheries: Importance, vulnerability and deficient knowledge. In S. Garcia, J. Rice, & A. Charles (Eds.), *Governance for marine fisheries and biodiversity conservation: Interaction and coevolution* (pp. 1–21). London: Wiley-Blackwell. Chapter 22.

Kooiman, J. (2003). *Governing as governance*. London: Sage.

Korda, R., Hills, J., & Gray, T. (2008). Fishery decline in Utila: Disentangling the web of governance. *Marine Policy, 32*(6), 968–979.

Kosamu, I. (2015). Conditions for sustainability of small-scale fisheries in developing countries. *Fisheries Research, 161*, 365–373.

Kraan, M. (2011). More than income alone: The anlo-ewe beach seine fishery in Ghana. In S. Jentoft & A. Eide (Eds.), *Poverty mosaics: Realities and prospects in small-scale fisheries* (pp. 147–172). Dordrecht: Springer.

Krogseng, K. (2016). *The decline of small-scale fisheries: A case study of Blekinge, Sweden*, Masters programme in rural development and natural resource management. Uppsala: Swedish University of Agricultural Sciences.

Krueger, R., & Casey, M. (2009). *Focus groups: A practical guide to applied research*. Thousand Oaks, CA: Sage.

Kurien, J. (2011). Foreword. In S. Jentoft & A. Eide (Eds.), *Poverty mosaics: Realities and prospects in small-scale fisheries* (pp. v–ix). Dordrecht: Springer.

Kvale, S. (1996). *Interviews: An introduction to qualitative research interviewing*. London: Sage.

Lalancette, A. (2017). Creeping in? Neoliberalism, indigenous realities and tropical rock lobster (*kaiar*) management in Torres Strait, Australia. *Marine Policy, 80*, 47–59.

Lam, M., & Pauly, D. (2010). Who is right to fish? Evolving a social contract for ethical fisheries. *Ecology and Society, 15*(3), 16.

Langdon, S. (2015). Foregone harvests and neoliberal policies: Creating opportunities for rural, small-scale, community-based fisheries in southern Alaskan coastal villages. *Marine Policy, 61*, 347–355.

Lauck, T., Clark, C., Mangel, M., & Munro, G. (1998). Implementing the precautionary principle in fisheries management through marine reserves. *Ecological Applications, 8*(1), 72–78.

Lebel, L., Anderies, J., Campbell, B., Folke, C., Hatfield-Dodds, S., Hughes, T., & Wilson, J. (2006). Governance and the capacity to manage resilience in regional social-ecological systems. *Ecology and Society, 11*(1), 19.

Leis, M., Chuenpagdee, R., & Medeiros, R. (2019). Where small-scale fisheries meet conservation boundaries: MPA governance challenges in Southern Brazil. In S. Salas, M. Barragán-Paladines, & R. Chuenpagdee (Eds.), *Viability and sustainability of small-scale fisheries in Latin America and the Caribbean* (pp. 453–472). Cham: Springer.

LIFE [Low Impact Fishers of Europe]. (2016, December). Rights-based management and small scale fisheries in the EU: Human rights versus property rights. *A LIFE position paper on ITQs*. Retrieved from lifeplatform.eu/wp-content/uploads/2016/12/LIFE-Statement-on-ITQs.pdf

Linkie, D. (2016, May 19). Industry mistrusts IFCAs. *Fishing News*, p. 4.

Linkie, D. (2019a, October 17). First step to new inshore regime. *Fishing News*, p. 4.

Linkie, D. (2019b, November 24). Time to change inshore management. *Fishing News*, p. 4.

Linkie, D. (2019c, December 12). New shellfish group a welcome move. *Fishing News*, p. 4.

LiPuma, E., & Meltzoff, S. (1997). The crosscurrents of ethnicity and class in the construction of public policy. *American Ethnologist, 24*(1), 114–131.

Lloret, J., Cowx, I., Cabral, H., et al. (2018). Small-scale coastal fisheries in European Seas are not what they were: Ecological, social, and economic changes. *Marine Policy, 98*, 178–186.

Locality. (2018). *People power: Findings from the commission on the future of localism*. London: Localism.

Longo, S., & Clark, B. (2012). The commodification of Bluefin tuna: The historical transformation of the Mediterranean fishery. *Journal of Agrarian Change, 2*(2–3), 204–226.

Lowther, J., & Rodwell, L. (2013). IFCAs: Stakeholder perceptions of roles, and legal impact. *Environmental Law Review, 15*(1), 1–14.

Macfadyen, G., Salz, P., & Cappell, R. (2011). Characteristics of small-scale coastal fisheries in Europe. *Directorate-General for Internal Policies*. Policy Department B: Structural and Cohesion Policies - Fisheries. Brussels: European Parliament. 156pp.

MacKenzie, B., & Schiedek, D. (2007). Daily ocean monitoring since the 1860s shows record warming of northern European seas. *Global Change Biology, 13*(7), 1335–1347.

Mackinson, S., & Wilson, D. C. K. (2014). Building bridges among scientists and fishermen with participatory action research. In J. Urquhart, T. Acott, D. Symes, & M. Zhao (Eds.), *Social issues in sustainable fisheries management* (pp. 121–139). Dordrecht: Springer.

MAF [Maritime Affairs and Fisheries]. (2016, October). What is the future for the small-scale coastal fleet? *Maritime Affairs and Fisheries Online Magazine*.

Mangi, S., Kupschus, S., Mackinson, S., Rodmell, D., Lee, A., Bourke, E., Rossiter, T., Masters, J., Hetherington, S., Catchpole, T., & Righton, D. (2018). Progress in designing and delivering effective fishing industry-science data collection in the UK. *Fish and Fisheries, 19*, 622–642.

Mansfield, B. (2004). Neoliberalism in the oceans: Rationalization, property rights, and the commons question. *Geoforum, 35*(3), 313–326.

Marschke, M., & Berkes, F. (2006). Exploring strategies that build livelihood resilience: A case from Cambodia. *Ecology and Society, 11*(1), 17.

Marshall, N., & Marshall, P. (2007). Conceptualizing and operationalizing social resilience within commercial fisheries in Northern Australia. *Ecology and Society, 12*(1), 14.

Mascia, M., Claus, C., & Naidoo, R. (2010). Impacts of marine protected areas on fishing communities. *Conservation Biology, 24*(5), 1424–1429.

Masozera, M., Bailey, M., & Kerchner, C. (2007). Distribution of impacts of natural disasters across income groups: A case study of New Orleans. *Ecological Economics, 63*(2–3), 299–306.

Mathew, S. (2003). Small-scale fisheries perspectives on an ecosystem-based approach to fisheries management. In M. Sinclair & G. Validimarsson (Eds.), *Responsible fisheries in the marine ecosystem* (pp. 47–63). Rome: CABI Publishing.

Maynou, F., Recasens, L., & Lombarte, A. (2011). Fishing tactics dynamics of a Mediterranean small-scale coastal fishery. *Aquatic Living Resources, 24*, 149–159.

McClanahan, T., Castilla, J., White, A., & Defeo, O. (2009). Healing small-scale fisheries by facilitating complex socio-ecological systems. *Reviews in Fish Biology and Fisheries, 19*, 33–47.

McConney, P., & Charles, A. (2012). Managing small-scale fisheries: Moving towards people-centred perspectives. In R. Grafton, R. Hilborn, D. Squires, M. Tait, & M. Williams (Eds.), *Handbook of marine fisheries conservation and management*. Oxford: Oxford University Press.

McConney, P., Medeiros, R., & Pena, M. (Eds.). (2014). *Enhancing stewardship in small-scale fisheries: Practice and perspectives*. TBTI: CERMES Technical Report No. 73.

McGoodwin, J. (1990). *Crisis in the world's fisheries: People, problems, and policies*. Stanford, CA: Stanford University Press.

McKibben, B. (2010). *Earth*. New York: Henry Holt.

Meadows, D., Meadows, D., Randers, J., & Behrens, W. (1972). *The limits to growth*. New York: Universe Books.

Meltzoff, S., & Schull, J. (1999). Miskito ethnic struggle over land and lobster: Conserving culture and resources on Corn Island. *Culture and Agriculture, 21*(3), 10–18.

Metzner, R., & Ward, J. (2002). *Report of the expert consultation on catalysing the transition away from overcapacity in marine capture fisheries*. FAO Fisheries Report. Rome: FAO.

Meyer, K. (2017). International business in an era of anti-globalization. *Multinational Business Review, 25*(2), 78–90.

MFA [Marine and Fisheries Agency]. (2009). *UK Sea fisheries statistics, 2008: Overview of the UK fishing industry* (pp. 78–90). London: Marine and Fisheries Agency Fisheries Statistics Unit.

Miller, K., & Fluharty, D. (1992). El Niño and variability in the north eastern Pacific salmon fishery: Implications for coping with climate change. In M. Glantz (Ed.), *Climate variability, climate change and fisheries* (pp. 49–88). Cambridge: Cambridge University Press.

Mills, D., Béné, C., Ovie, S., Tafida, A., Sinaba, F., Kodio, A., & Lemoalle, J. (2011a). Vulnerability in African small-scale fishing communities. *Journal of International Development, 23*(2), 308–313.

Mills, D., Westlund, L., de Graaf, G., Kura, Y., Willman, R., & Keleher, K. (2011b). Under-reported and under-valued: Small-scale fisheries in the developing world. In R. Pomeroy & N. Andrew (Eds.), *Small-scale fisheries management: Frameworks and approaches for the developing world* (pp. 1–15). Oxfordshire: CAB International.

MMO [Marine Management Organisation]. (2016a). *Sea fisheries statistics 2015*. Newcastle.

MMO [Marine Management Organisation]. (2016b). *Appointment of inshore fisheries and conservation authority members by the Marine Management Organisation. Information for candidates*. Newcastle.

MMO [Marine Management Organisation]. (2018). *Statistics at MMO*. Newcastle.

Mohan, G., & Stokke, K. (2000). Participatory development and empowerment: The dangers of localism. *Third World Quarterly, 21*(2), 247–268.

Monnereau, I., & McConney, P. (2015). Governability of small-scale lobster fisheries in the Caribbean. In S. Jentoft & R. Chuenpagdee (Eds.), *Interactive governance for small-scale fisheries* (pp. 223–241). Cham: Springer.

Morgan, R. (2013). Exploring how fishermen respond to the challenges facing the fishing industry: A study of diversification and multiple-job holding in the English Channel fishery (Doctoral dissertation, University of Portsmouth).

Morgan, R. (2016). Exploring how fishermen respond to the challenges facing the fishing industry: A case study of diversification in the English Channel fishery. *Regional Studies, 50*(1), 1755–1768.

Morgan, D., & Spanish, M. (1984). Focus groups: A new tool for qualitative research. *Qualitative Sociology, 7*(3), 253–270.

Morton, J. (2007). The impact of climate change on smallholder and subsistence agriculture. *Proceedings of the National Academy of Sciences, 104*(50), 19680–19685.

Morzaria-Luna, H., Turk-Boyer, P., & Moreno-Baez, M. (2014). Social indicators of vulnerability for fishing communities in the Northern Gulf of California, Mexico: Implications for climate change. *Marine Policy, 45*, 182–193.

Mowbray, D. (2017). *Understanding resilience*. Cheltenham: Management Advisory Service.

Msomphora, M. (2015). Stakeholder participation and satisfaction in the process of developing management plans: The case of Scottish Inshore Fisheries Groups. *Ocean and Coastal Management, 116*, 491–503.

Nadel-Klein, J. (2000). Granny baited the lines: Perpetual crisis and the changing role of women in Scottish fishing communities. *Women's Studies International Forum, 23*(3), 363–372.

Nayak, P., & Berkes, F. (2010). Whose marginalisation? Politics around environmental injustices in India's Chilika lagoon. *Local Environment, 15*(6), 553–567.

Nayak, P., & Berkes, F. (2019). Interplay between local and global: Change processes and small-scale fisheries. In R. Chuenpagdee & S. Jentoft (Eds.), *Transdisciplinarity for small-scale fisheries governance: Analysis and practice* (pp. 203–220). Cham: Springer.

NEIFCA [North Eastern Inshore Fisheries Authority]. (2015/16). *Annual plan 2015/16*. Bridlington: Author.

NFFO [National Federation of Fishermen's Organisations]. (2019a, April). *Fixed quota allocations and sustainable fishing*. York: Author.

NFFO [National Federation of Fishermen's Organisations]. (2019b, October). *Fishing quota allocation: Developing a new approach for allocating additional fishing quota in England*. York: Author.

Nguyen, K., & Flaaten, O. (2011). Facilitating change: A Mekong Vietnamese small-scale fishing community. In S. Jentoft & A. Eide (Eds.), *Poverty mosaics: Realities and prospects in small-scale fisheries* (pp. 335–337). Dordrecht: Springer.

Nimmo, F., & Southall, T. (2012). *Project inshore*. Inverness: Food Certification International.

Norberg-Hodge, H. (2001). Local lifeline: Rejecting globalization—Embracing localization. In V. Bennholdt-Thomsen, N. Faraclas, & C. von Werlhof (Eds.), *There is an alternative: Subsistence and world-wide resistance to corporate globalization* (pp. 178–188). London: Zed Books.

Nozick, R. (1974). *Anarchy, state and utopia*. New York: Basic Books.

Nunoo, F., Eggleston, D., & Vanderpuye, C. (2006). Abundance, biomass and species composition of nearshore fish assemblages in Ghana, West Africa. *African Journal of Marine Science, 28*(3–4), 689–696.

NUTFA [New Under Ten Fishermen's Association]. (2014). *Fisheries management*. Ivybridge: Author.

NUTFA [New Under Ten Fishermen's Association]. (2019, October 24). NUTFA: Inshore management needs wholesale change. *Fishing News*, pp. 8–9.

Nuttall, M. (2000). Crisis, risk and deskilment in North-East Scotland's fishing industry. In D. Symes (Ed.), *Fisheries dependent regions* (pp. 106–115). Oxford: Blackwell Science.

Oliver, T. (2014, November 28). A voice at the top for small scale fishermen. *Fishing News*, p. 5.

Oliver, T. (2015a, February 6). MMO under fire at select committee review. *Fishing News*, p. 2.

Oliver, T. (2015b, May 29). New FPO for under-10 and non-sector fishermen. *Fishing News*, p. 2.

Oliver, T. (2016, May 19). Not enough industry reps on IFCA boards. *Fishing News*, p. 2.

Oliver, T. (2019a, July 4). Fisheries bill change aims to end FQAs. *Fishing News*, p. 2.

Oliver, T. (2019b, August 29). Major conference on inshore future. *Fishing News*, p. 2.

Oliver, T. (2019c, November 14). Landmark conference kick-starts new era for inshore sector. *Fishing News*, p. 10.

Oliver, T. (2019d, December 5). Fishermen "drivers" of landmark inshore conference. *Fishing News*, p. 7.

Oliver, T. (2019e, December 12). Industry move to highlight shellfish management. *Fishing News*, p. 2.

Oliver, T. (2019f, December 19). LIFE: Right historic wrongs against small-scale fishermen. *Fishing News*, p. 5.

Oliver, T. (2019g, December 19). Inshore leader slams discard ban. *Fishing News*, p. 5.

Onyango, P. (2011). Occupation of last resort? Small-scale fishing in Lake Victoria, Tanzania. In S. Jentoft & A. Eide (Eds.), *Poverty mosaics: Realities and prospects in small-scale fisheries* (pp. 97–124). Dordrecht: Springer.

Ota, Y., & Just, R. (2008). Fleet sizes, fishing effort and the 'hidden' factors behind statistics: An anthropological study of small-scale fisheries in UK. *Marine Policy, 32*(3), 301–308.

Paladines, M. (2015). Two rules for the same fish: Small-scale fisheries governance in mainland Ecuador and Galapagos Islands. In S. Jentoft & R. Chuenpagdee (Eds.), *Interactive governance for small-scale fisheries* (pp. 157–178). Cham: Springer.

Parks, J. (2011). Adaptive management in small-scale fisheries: A practical approach. In R. Pomeroy & N. Andrew (Eds.), *Small scale fisheries management: Frameworks and approaches for the developing world* (pp. 93–114). Oxfordshire: CAB International.

Pascual-Fernández, J., Pita, C., Josupeit, H., Said, A., & Rodrigues, J. (2019). Markets, distribution and value chains in small-scale fisheries: A special focus on Europe. In R. Chuenpagdee & S. Jentoft (Eds.), *Transdisciplinarity for small-scale fisheries governance: Analysis and practice* (pp. 141–162). Cham: Springer.

Pauly, D. (2006). Major trends in small-scale marine fisheries, with emphasis on developing countries, and some implications for the social sciences. *MAST, 4*(2), 7–22.

Pauly, D. (2011). Foreword. In R. Chuenpagdee (Ed.), *World small-scale fisheries: Contemporary visions* (pp. 15–17). Delft: Eburon.

Pauly, D. (2018). A vision for marine fisheries in a global blue economy. *Marine Policy, 87*, 371–374.

Pearson, L., & Pearson, C. (2012). Societal collapse or transformation and resilience. *Proceedings of the National Academy of Sciences, 109*(30), 2030–2031.

Pelling, M. (1998). Participation, social capital and vulnerability to urban flooding in Guyana. *Journal of International Development, 10*(4), 469–486.

Pena, M., & McConney, P. (2014). Socio-economic monitoring for coastal management (SocMon): Application of a participatory monitoring tool to small-scale fisheries. In P. McConney, R. Medeiros, & M. Pena (Eds.), *Enhancing stewardship in small-scale fisheries: Practice and perspectives*. TBTI: CERMES Technical Report No. 73, pp. 49–56.

Penca, J. (2019). Transnational localism: Empowerment through standard setting in small-scale fisheries. *Transnational Environmental Law, 8*(1), 143–165.
Percy, J. (2014). Interview with Jeremy Percy, executive director of LIFE. *European Maritime Affairs and Fisheries, 65*(October), 6–7.
Percy, J. (2015). Delegate notes. *Coastal futures, 2015.*
Percy, J. (2016, June 7). *Small-scale fishing: A solution rather than a problem.* Life Platform, EU.
Percy, J. (2018). *Why the government's fisheries White Paper could mean a bad deal for local, low impact fishing.* Ivybridge: NUTFA.
Phillipson, J. (2002). *Widening the net: Prospects for fisheries co-management.* Newcastle upon Tyne: Newcastle University, Centre for Rural Economy.
Phillipson, J., & Symes, D. (2010). Recontextualising inshore fisheries: The changing face of British inshore fisheries management. *Marine Policy, 34*(6), 1207–1214.
Phillipson, J., & Symes, D. (2015). Finding a middle way to develop Europe's fisheries dependent areas: The role of Fisheries Local Action Groups. *Sociologia Ruralis, 55*(3), 343–359.
Phillipson, J., & Symes, D. (2018). 'A sea of troubles': Brexit and the fisheries question. *Marine Policy, 90*, 168–173.
Pieraccini, M., & Cardwell, E. (2016). Towards deliberative and pragmatic co-management: A comparison between inshore fisheries authorities in England and Scotland. *Environmental Politics, 25*(4), 729–748.
Pinkerton, E. (2003). Toward specificity in complexity: Understanding co-management from a social science perspective. In D. Wilson, J. Nielsen, & P. Degnbol (Eds.), *The fisheries co-management experience: Accomplishments, challenges and prospects* (pp. 61–77). Dordrecht: Kluwer.
Pinkerton, E. (2015). The role of moral economy in two British Columbia fisheries: Confronting neoliberal policies. *Marine Policy, 61*, 410–419.
Pinkerton, E. (2017). Hegemony and resistance: Disturbing patterns and hopeful signs in the impact of neoliberal policies on small-scale fisheries around the world. *Marine Policy, 80*, 1–9.
Pinkerton, E. (2019). Strategies and policies supporting small-scale fishers' access and conservation rights in a neoliberal world. In R. Chuenpagdee & S. Jentoft (Eds.), *Transdisciplinarity for small-scale fisheries governance: Analysis and practice* (pp. 241–261). Cham: Springer.
Pinkerton, E., & Davis, R. (2015). Neoliberalism and the politics of enclosure in North American small-scale fisheries. *Marine Policy, 61*, 303–312.
Pinkerton, E., & Edwards, D. (2009). The elephant in the room: The hidden costs of leasing individual transferable fishing quotas. *Marine Policy, 33*, 707–713.
Pita, C., Pierce, G., & Theodossiou, I. (2010). Stakeholders' participation in the fisheries management decision-making process: Fishers' perceptions of participation. *Marine Policy, 34*(5), 1093–1102.
Pita, C., Pereira, J., Lourenço, S., Sonderblohm, C., & Pierce, G. (2015). The traditional small-scale octopus fishery in Portugal: Framing its governability. In S. Jentoft & R. Chuenpagdee (Eds.), *Interactive governance for small-scale fisheries* (pp. 117–132). Cham: Springer.
Pitchon, P. (1997, April). *Globalization versus localization.* Share International.
Pollnac, R. (1979). The structure of job satisfaction among New England fishermen. In J. Acheson (Ed.), *Essays on social and cultural aspects of New England fisheries: Implications for management, 1980 final report, volume II* (p. 255). Orono: The University of Maine.
Pollnac, R., & Poggie, J. (2006). Job satisfaction in the fishery in two southeast Alaskan towns. *Human Organization, 65*, 329–339.
Pollnac, R., & Poggie, J. (2008). Happiness, well-being and psychocultural adaptation to the stresses associated with marine fishing. *Human Ecology Review, 15*(2), 194–200.
Pomeroy, R. (2012). Managing overcapacity in small-scale fisheries in Southeast Asia. *Marine Policy, 36*, 520–527.
Porter, M. (1989). *Not drowning but waving: Reading nineteenth century whaling women's diaries* (Studies in sexual politics) (Vol. 28, pp. 1–77). Manchester: Sociology Department, University of Manchester.

Pörtner, H., Karl, D., Boyd, P., Cheung, W., Lluch-Cota, S., Nojiri, Y., & Armstrong, C. (2014). 'Ocean systems' in IPCC Climate change 2014: Impacts, adaptation, and vulnerability. Part A: Global and sectoral aspects. In *Contribution of working group II to the fifth assessment report of the intergovernmental panel on climate change*. Cambridge: Cambridge University Press, pp. 411–484.

Prosperi, P., Kirwan, J., Maye, D., Bartoloni, F., Vergamini, D., & Brunori, G. (2019). Adaptation strategies of small-scale fisheries within changing market and regulatory conditions in the EU. *Marine Policy, 100*, 316–323.

Prowse, M. (2003). *Towards a clearer understanding of 'vulnerability' in relation to chronic poverty*. CPRC Working Paper No. 24, University of Manchester. Manchester: Chronic Poverty Research Centre.

Quist, L., & Nygren, A. (2015). Contested claims over space and identity between fishers and the oil industry in Mexico. *Geoforum, 63*, 44–54.

Ramirez-Luna, V., & Chuenpagdee, R. (2019). Exclusive fishing zone for small-scale fisheries in Northern Chocó, Colombia: Pre-and post-implementation. In S. Salas, M. Barragán-Paladines, & R. Chuenpagdee (Eds.), *Viability and sustainability of small-scale fisheries in Latin America and the Caribbean* (pp. 197–217). Cham: Springer.

Ratner, B., & Allison, E. (2012). Wealth, rights, and resilience: An agenda for governance reform in small-scale fisheries. *Development Policy Review, 30*(4), 371–398.

Reed, M., Courtney, P., Urquhart, J., & Ross, N. (2013). Beyond fish as commodities: Understanding the socio-cultural role of inshore fisheries in England. *Marine Policy, 37*, 62–68.

Richardson, E., Kaiser, M., & Edwards-Jones, G. (2005). Variation in fishers' attitudes within an inshore fishery: Implications for management. *Environmental Conservation, 32*(3), 213–225.

Robson, C. (2011). *Real world research* (3rd ed.). Chichester: Wiley.

Rodwell, L., Lowther, J., Hunter, C., & Mangi, S. (2014). Fisheries co-management in a new era of marine policy in the UK: A preliminary assessment of stakeholder perceptions. *Marine Policy, 45*, 279–286.

Ross, N. (2013). Exploring concepts of fisheries 'dependency' and 'community' in Scotland. *Marine Policy, 37*, 55–61.

Rothschild, B., Chen, C., & Lough, R. (2005). Managing fish stocks under climate uncertainty. *ICES Journal of Marine Science, 62*(7), 1531–1541.

Ruddle, K., & Davis, A. (2013). Human rights and neo-liberalism in small-scale fisheries: Conjoined priorities and processes. *Marine Policy, 39*, 87–93.

Sabau, G., & van Zyll de Jong, M. (2015). From unjust economic growth to sustainable fisheries in Newfoundland: The true costs of closing the inshore fishery. *Marine Policy, 61*, 376–389.

Salas, S., & Pitcher, T. (1999). Fishing strategies of small-scale lobster fishermen in Yucatan, Mexico. *Revista de Ciencias del Mar, 16*, 30–38.

Salas, S., Chuenpagdee, R., Seijo, J., & Charles, A. (2007). Challenges in the assessment and management of small-scale fisheries in Latin America and the Caribbean. *Fisheries Research, 87*(1), 5–16.

Salas, S., Bjørkan, M., Bobadilla, F., & Cabrera, M. (2011). Addressing vulnerability: Coping strategies of fishing communities in Yucatan, Mexico. In S. Jentoft & A. Eide (Eds.), *Poverty mosaics: Realities and prospects in small-scale fisheries* (pp. 195–220). Dordrecht: Springer.

Salas, S., Barragán-Paladines, M., & Chuenpagdee, R. (Eds.). (2019). *Viability and sustainability of small-scale fisheries in Latin America and the Caribbean*. Cham: Springer.

Salmi, P. (2005). Rural pluriactivity as a coping strategy in small-scale fisheries. *Sociologia Ruralis, 45*(1–2), 22–36.

Salmi, P. (2015). Constraints and opportunities for small-scale fishing livelihoods in a post-productivist coastal setting. *Sociologia Ruralis, 55*(3), 258–274.

Sathyapalan, J., & George, S. (2015). Governability challenges in sustaining small-scale fisheries in an urban context: A study of Cochin Backwaters, India. In S. Jentoft & R. Chuenpagdee (Eds.), *Interactive governance for small-scale fisheries* (pp. 85–99). Cham: Springer.

Scholtens, J. (2016). The elusive quest for access and collective action: North Sri Lankan fishers' thwarted struggles against a foreign trawler fleet. *International Journal of the Commons, 10*(2), 929–952.

Scholtens, J., Song, A., Stephen, J., Chavez, C., Bavinck, M., & Sowman, M. (2019). Transdisciplinary engagement to address transboundary challenges for small-scale fishers. In R. Chuenpagdee & S. Jentoft (Eds.), *Transdisciplinarity for small-scale fisheries governance: Analysis and practice* (pp. 321–340). Cham: Springer.

Schumacher, E. (1973). *Small is beautiful: A study of economics as if people mattered.* London: Harper Collins.

Schumann, S., & Macinko, S. (2007). Subsistence in coastal fisheries policy: What's in a word? *Marine Policy, 31*(6), 706–718.

Schwarz, A., Béné, C., Bennett, G., Boso, D., Hilly, Z., Paul, C., & Andrew, N. (2011). Vulnerability and resilience of remote rural communities to shocks and global changes: Empirical analysis from Solomon Islands. *Global Environmental Change, 21*(3), 1128–1140.

Seidman, I. (1991). *Interviewing as qualitative research: A guide for researchers in education and the social sciences*. New York: Teachers College Press.

Seixas, C., Davidson-Hunt, I., Kalikoski, D., Davy, B., Berkes, F., de Castro, F., Medeirminte-Vera, C., & Arujo, L. (2019). Collaborative coastal management in Brazil: Advancements, challenges and opportunities. In S. Salas, M. Barragán-Paladines, & R. Chuenpagdee (Eds.), *Viability and sustainability of small-scale fisheries in Latin America and the Caribbean* (pp. 425–451). Cham: Springer.

Sen, A. (2000). *Development as freedom*. New York: Anchor Books.

Sharma, C. (2004). The impact of fisheries development and globalization processes on women of fishing communities in the Asian region. *Asia Pacific Research Network, 8*, 1–12.

Skaptadóttir, U. (2000). Women coping with change in an Icelandic fishing community: A case study. *Women's Studies International Forum, 23*(3), 311–321.

Smith, H. (2000). The industrialisation of the world ocean. *Ocean and Coastal Management, 43*, 11–28.

Smith, H. (2013). The regional development and management of fisheries: The UK case. *Marine Policy, 37*, 11–19.

Smith, H., & Basurto, X. (2019). Defining small-scale fisheries and examining the role of science in shaping perceptions of who and what counts: A systematic review. *Frontiers in Marine Science, 6*, 236.

Smith, S., Battista, W., Sarto, N., Fujita, R., Stetten, D., Karasik, R., & Burden, M. (2019). A framework for allocating fishing rights in small-scale fisheries. *Ocean and Coastal Management, 177*, 52–63.

Snape, D., & Spencer, L. (2003). Introduction to research method. In J. Ritchie & J. Lewis (Eds.), *Qualitative research practice: A guide for social science students and researchers*. London: Sage.

Soliman, A. (2015). Alaska's community quota entities programme for halibut and sablefish: Between governability challenges and opportunities. In S. Jentoft & R. Chuenpagdee (Eds.), *Interactive governance for small-scale fisheries* (pp. 299–318). Cham: Springer.

Song, A., & Soliman, A. (2019). Situating human rights in the context of fishing rights—Contributions and contradictions. *Marine Policy, 103*, 19–26.

Song, A., Bodwitch, H., & Scholtens, J. (2018). Why marginality persists in a governable fishery—The case of New Zealand. *Maritime Studies, 17*, 285–293.

Sowman, M. (2006). Subsistence and small-scale fisheries in South Africa: A ten-year review. *Marine Policy, 30*(1), 60–73.

Sowman, M., & Cardoso, P. (2010). Small-scale fisheries and food security strategies in countries in the Benguela Current Large Marine Ecosystem (BCLME) region: Angola, Namibia and South Africa. *Marine Policy, 34*, 1163–1170.

Sowman, M., Sunde, J., Raemaekers, S., & Schultz, O. (2014). Fishing for equality: Policy for poverty alleviation for South Africa's small-scale fisheries. *Marine Policy, 46*, 31–42.

Stanton, J. (2014). The big society and community development: Neighbourhood planning under the Localism Act. *Environmental Law Review, 16*(4), 262–276.

Start, D., & Johnson, C. (2004). *Livelihood options? The political economy of access, opportunity and diversification*. London: Overseas Development Institute.

Stewart, J., & Callagher, P. (2011). Quota concentration in the New Zealand fishery: Annual catch entitlement and the small fisher. *Marine Policy, 35*(5), 631–646.

Stewart, J., Walsh, K., & Moodie, B. (2006). The demise of the small fisher? A profile of exiters from the New Zealand fishery. *Marine Policy, 30*(4), 328–340.

Stiglitz, J. (2003). The promise of global institutions. In D. Held & A. McGrew (Eds.), *The global transformations reader: An introduction to the globalization debate* (pp. 477–481). Cambridge: Polity Press.

Stobberup, K., Garza Gil, M., Stirnemann-Relot, A., Rigaud, A., Franceschelli, N., & Blomeyer, R. (2017). *Small-scale fisheries and "blue growth" in the EU. European Parliament*. Brussels: Policy Department for Structural and Cohesion Policies.

Sturges, J., & Hanrahan, K. (2004). Comparing telephone and face-to-face qualitative interviewing: A research note. *Qualitative Research, 4*(1), 107–118.

Sumaila, U. (2010). A cautionary note on individual transferable quotas. *Ecology and Society, 15*(3), 36.

Suuronen, P., Chopin, F., Glass, C., Løkkeborg, S., Matsushita, Y., Queirolo, D., & Rihan, D. (2012). Low impact and fuel efficient fishing—Looking beyond the horizon. *Fisheries Research, 119-120*, 135–146.

Symes, D. (1992). The common fisheries policy and UK quota management. *Ocean and Coastal Management, 18*(2–4), 319–338.

Symes, D. (2000). Fisheries dependent regions: Scoping the problem. In D. Symes (Ed.), *Fisheries dependent regions* (pp. 3–14). Oxford: Blackwell Science.

Symes, D. (2001). Inshore fisheries in Europe at the turn of the century. In D. Symes & J. Phillipson (Eds.), *Inshore fisheries management. Reviews: Methods and technologies in fish biology* (pp. 3–23). Dordrecht: Springer.

Symes, D. (2002). *Inshore fisheries management in England and Wales: Facing up to the challenges of the 21st century*. Peterborough: English Nature Research Report No. 448.

Symes, D. (2009). Regionalising the Common Fisheries Policy: What kind of institutional solution? In *Regionalisation of the EU's Common Fisheries Policy*. Copenhagen: Nordic Council of Ministers' Conference.

Symes, D. (2014). Finding solutions: Resilience theory and Europe's small-scale fisheries. In J. Urquhart, T. Acott, & D. Symes (Eds.), *Social issues in sustainable fisheries management* (MARE publication series 9). Dordrecht: Springer.

Symes, D., & Phillipson, J. (1997). Inshore fisheries management in the UK: Sea Fisheries Committees and the challenge of marine environmental management. *Marine Policy, 21*(3), 207–224.

Symes, D., & Phillipson, J. (2009). Whatever became of social objectives in fisheries policy? *Fisheries Research, 95*(1), 1–5.

Symes, D., Phillipson, J., & Salmi, P. (2015). Europe's coastal fisheries: Instability and the impacts of fisheries policy. *Sociologia Ruralis, 55*(3), 245–257.

Taylor, R. (1962). Fatalism. *Philosophical Review, 71*(1), 56–66.

TBTI [Too Big To Ignore]. (2017). *TBTI symposium on European small-scale fisheries and global linkages, June 29–July 1, 2016*, Santa Cruz de Tenerife, Spain.

Teddlie, C., & Tashakkori, A. (2009). *Foundations of mixed methods research: Integrating quantitative and qualitative approaches in the social and behavioral sciences*. Thousand Oaks, CA: Sage.

Tep, R., & Fauchald, N. (2007, February). How to eat like a locavore. *Food and Wine*.

Thiessen, V., Davis, A., & Jentoft, S. (1992). The veiled crew: An exploratory study of wives' reported and desired contributions to coastal fisheries enterprises in northern Norway and Nova Scotia. *Human Organization, 51*, 342–352.

Thomas, H., Macsharry, B., Morgan, L., Kingston, N., Moffitt, R., Stanwell-Smith, D., & Wood, L. (2014). Evaluating official marine protected area coverage for Aichi Target 11: Appraising the data and methods that define our progress. *Aquatic Conservation: Marine and Freshwater Ecosystems, 24*(S2), 8–23.

Thompson, P., Wailey, T., & Lummis, T. (1983). *Living the fishing*. London: Routledge.

Thornton, T., & Hebert, J. (2015). Neoliberal and neo-communal herring fisheries in Southeast Alaska: Reframing sustainability in marine ecosystems. *Marine Policy, 61*, 366–375.

Thorpe, A., Andrew, N., & Allison, E. (2007). Fisheries and poverty reduction. *CAB Reviews: Perspectives in Agriculture, Veterinary Science, Nutrition and Natural Resources, 2*(85), 2–12.

Thorpe, A., Bavinck, M., & Coulthard, S. (2011). Tracking the debate around marine protected areas: Key issues and the BEG framework. *Environmental Management, 47*(4), 546–563.

Tolley, B., & Hall-Arber, M. (2015). Tipping the scale away from privatization and towards community-based fisheries: Policy and market alternatives in New England. *Marine Policy, 61*, 401–409.

Tompkins, E., & Adger, W. (2004). Does adaptive management of natural resources enhance resilience to climate change? *Ecology and Society, 9*(2), 10.

Trimble, M. (2014). Multi-stakeholder participatory research: Opportunities and challenges in coastal Uruguay. In P. McConney, R. Medeiros, & M. Pena (Eds.), *Enhancing stewardship in small-scale fisheries: Practice and perspectives*. TBTI: CERMES Technical Report No. 73, pp. 98–105.

Trimble, M., & Johnson, D. (2013). Artisanal fishing as an undesirable way of life? The implications for governance of fishers' wellbeing aspirations in coastal Uruguay and south-eastern Brazil. *Marine Policy, 37*, 37–44.

Turner, B., Kasperson, R., Matson, P., McCarthy, J., Corell, R., Christensen, L., & Polsky, C. (2003). A framework for vulnerability analysis in sustainability science. *Proceedings of the National Academy of Sciences, 100*(14), 8074–8079.

Tvedten, I., & Hersoug, B. (1992). *Fishing for development: Small-scale fisheries in Africa*. Uppsala: Scandinavian Institute of African Studies.

UNDP [United Nations Development Programme]. (2003). Patterns of global inequality. In D. Held & A. McGrew (Eds.), *The global transformations reader: An introduction to the globalization debate* (pp. 423–429). Cambridge: Polity Press.

UNDP [United Nations Development Programme]. (2005). *Human development reports*. New York: United Nations.

Urquhart, J., & Acott, T. (2013). Constructing 'the Stade': Fishers' and non-fishers' identity and place attachment in Hastings, south-east England. *Marine Policy, 37*, 45–54.

Urquhart, J., & Acott, T. (2014). A sense of place in cultural ecosystem services: The case of Cornish fishing communities. *Society and Natural Resources, 27*(1), 3–19.

Urquhart, J., Acott, T., Reed, M., & Courtney, P. (2011). Setting an agenda for social science research in fisheries policy in Northern Europe. *Fisheries Research, 108*(2–3), 240–247.

Urquhart, J., Acott, T., & Zhao, M. (2013). Introduction: Social and cultural impacts of marine fisheries. *Marine Policy, 37*, 1–2.

Van de Walle, G., Da Silva, S., O'Hara, E., & Soto, P. (2015). Achieving sustainable development of local fishing interests: The case of Pays d'Auray flag. *Sociologia Ruralis, 55*(3), 360–377.

Van Ginkel, R. (2001). Inshore fishermen: Cultural dimensions of a maritime occupation. In D. Symes & J. Phillipson (Eds.), *Inshore fisheries management. Reviews: Methods and technologies in fish biology* (pp. 177–193). Dordrecht: Springer.

Vaughan, M., & Vitousek, P. (2013). Mahele: Sustaining communities through small-scale inshore fishery catch and sharing networks. *Pacific Science, 67*(3), 329–344.

Veiga, P., Pita, C., Rangel, M., Goncalves, J., Campos, A., Fernandez, P., Sala, A., Virgili, M., Luchetti, A., & Bricic, J. (2016). The EU landing obligation and European small-scale fisheries: What are the odds for success? *Marine Policy, 64*, 64–71.

Villasante, S., Garcia-Negro, M., Gonzales-Laxe, F., & Rodriguez, G. (2011). Overfishing and the Common Fisheries Policy: [un]successful results from TAC regulations? *Fish and Fisheries, 12*(1), 34–50.

Walker, B., Gunderson, L., Kinzig, A., Folke, C., Carpenter, S., & Schultz, L. (2006). A handful of heuristics and some propositions for understanding resilience in social-ecological systems. *Ecology and Society, 11*(1), 13.

Ward, J., Pascoe, S., Kirkley, J., Metzner, R., & Greboval, D. (2004). *Measuring and assessing capacity in fisheries: Basic concepts and management options*. Rome: FAO.

Warner, A. (2005). A brief history of the anti-globalization movement. *University of Miami International and Comparative Law Review, 12*, 237–268.

Weeratunge, N., Béné, C., Siriwardane, R., Charles, A., Johnson, D., Allison, E., Nayak, P., & Badjeck, M.-C. (2014). Small-scale fisheries through the wellbeing lens. *Fish and Fisheries, 15*, 255–279.

West, P., Igoe, J., & Brockington, D. (2006). Parks and peoples: The social impact of protected areas. *Annual Review of Anthropology, 35*, 251–277.

Weyl, O., & Weyl, M. (2001). *Proceedings of the Lake Malawi fisheries management symposium*, 4–9 June, Lilongwe, Malawi.

Willmann, R. (2015). Foreword. In S. Jentoft & R. Chuenpagdee (Eds.), *Interactive governance for small-scale fisheries. Global reflections* (pp. ix–xiii). Dordrecht: Springer.

Willmann, R., Franz, N., Fuentevilla, C., McInerney, T. F., & Westlund, L. (2017a). A human rights-based approach to securing small-scale fisheries: A quest for development as freedom. In S. Jentoft, R. Chuenpagdee, M. Barragan-Paladines, & N. Franz (Eds.), *The small-scale fisheries guidelines: Global implementation* (pp. 15–34). Dordrecht: Springer.

Willmann, R., Franz, N., Fuentevilla, C., McInerney, T. F., & Westlund, L. (2017b). A human rights-based approach in small-scale fisheries: Evolution and challenges in implementation. In S. Jentoft, R. Chuenpagdee, M. Barragan-Paladines, & N. Franz (Eds.), *The small-scale fisheries guidelines: Global implementation* (pp. 763–787). Dordrecht: Springer.

Wills, J. (2016). *Locating localism: Statecraft, citizenship and democracy*. Bristol: Policy Press.

Wilson, D., Nielsen, J., & Degnbol, P. (Eds.). (2003). *The fisheries co-management experience: Accomplishments, challenges and prospects*. Dordrecht: Kluwer.

Wilson, S., Pearson, L., Kashima, Y., Lusher, D., & Pearson, C. (2013). Separating adaptive maintenance (resilience) and transformative capacity of social-ecological systems. *Ecology and Society, 18*(1), 22.

Witter, A., & Stoll, J. (2017). Participation and resistance: Alternative seafood marketing in a neoliberal era. *Marine Policy, 80*, 130–140.

Wood, G. (2003). Staying secure, staying poor: The "Faustian Bargain". *World Development, 31*(3), 455–471.

Woods, N. (2003). Order, globalization and inequality in world politics. In D. Held & A. McGrew (Eds.), *The global transformations reader: An introduction to the globalization debate* (pp. 463–476). Cambridge: Polity Press.

World Bank. (2000). *World development report 2000/1*. Oxford: Oxford University Press.

World Bank/FAO/WorldFish. (2010). *The hidden harvests: The global contribution of capture fisheries*. Agriculture and Rural Development Department, Sustainable Development Network. Washington, DC: World Bank.

Yodanis, C. (2000). Constructing gender and occupational segregation: A study of women and work in fishing communities. *Qualitative Sociology, 23*, 267–290.

Zeller, D., & Pauly, D. (2005). Good news, bad news: Global fisheries discards are declining, but so are total catches. *Fish and Fisheries, 6*(2), 156–159.

Zhao, M., Tyzack, M., Anderson, R., & Onoakpovika, E. (2013). Women as visible and invisible workers in fisheries: A case study of Northern England. *Marine Policy, 37*(1), 69–76.

Index

© The Editor(s) (if applicable) and The Author(s), under exclusive license to Springer Nature Switzerland AG 2021
R. Korda et al., *Resilience in the English Small-Scale Fishery*, https://doi.org/10.1007/978-3-030-54245-0

The manufacturer's authorised representative in the EU is Springer Nature Customer Service Centre GmbH, Europaplatz 3, 69115 Heidelberg, Germany. If you have any concerns regarding our products, please contact ProductSafety@springernature.com

Printed and bound by CPI Group (UK) Ltd, Croydon, CR0 4YY
30/07/2026
02181626-0006